Managing the Local Climate

Praise for this book

Somaliland pastoralists are changing microclimates in 23 sites in the midst of degraded rangelands. Threatened by climate change, in their quest for more fodder, they protect thousands of hectares and proudly show grass and bush species that were thought to be extinct, and springs that produce water and attract wildlife, thought to be long gone. This book gives us great inspiration that on top of all these little victories against hunger, we can create a patchwork of improved microclimates across Somaliland – and in the near future, across the Horn of Africa.
Thomas Hoerz, Welthungerhilfe, Somaliland

The bottom up and proactive steps to manage local climates contained within this practical and well written book are the perfect resilience strategy. They provide a natural, positive way for people to thrive within uncertainty. This book is essential reading for everyone working in natural resource management.
Professor Nathanial Matthews, CEO of the Global Resilience Partnership

Both empirical as well as scientific evidence have confirmed the two-ways interaction of land and climate. This book adequately presents such proofs, and more importantly, describes how to manage the local climate as a new way to positively respond to climate change.
Yasir A. Mohamed, Ex minister of water resources, Sudan

Managing the Local Climate
How to positively affect the local climate and agroecological conditions with local interventions

Femke van Woesik,
Frank van Steenbergen,
Francesco Sambalino, Hugo Jan de Boer,
Jean Marc Pace Ricci
and Wim Bastiaanssen

Practical Action Publishing Ltd
25 Albert Street, Rugby,
Warwickshire, CV21 2SG, UK
www.practicalactionpublishing.com

© Femke van Woesik, Frank van Steenbergen, Francesco Sambalino, Hugo Jan de Boer, Jean Marc Pace Ricci and Wim Bastiaanssen, 2023

The moral right of the authors to be identified as authors of the work has been asserted under sections 77 and 78 of the Copyright Design and Patents Act 1988.

All rights reserved. No part of this publication may be reprinted or reproduced or utilized in any form or by any electronic, mechanical, or other means, now known or hereafter invented, including photocopying and recording, or in any information storage or retrieval system, without the written permission of the publishers.

Product or corporate names may be trademarks or registered trademarks, and are used only for identification and explanation without intent to infringe.

A catalogue record for this book is available from the British Library.

A catalogue record for this book has been requested from the Library of Congress.

ISBN 978-1-78853-225-9 Paperback
ISBN 978-1-78853-224-2 Hardback
ISBN 978-1-78853-226-6 Electronic book

Citation: van Woesik, F., van Steenbergen, F., Sambalino, F., de Boer, H.J., Pace Ricci, J.M. and Bastiaanssen, W. (2023) *Managing the Local Climate: How to positively affect the local climate and agroecological conditions with local interventions*, Rugby, UK: Practical Action Publishing <http://doi.org/10.3362/9781788532266>.

Since 1974, Practical Action Publishing has published and disseminated books and information in support of international development work throughout the world. Practical Action Publishing is a trading name of Practical Action Publishing Ltd (Company Reg. No. 1159018), the wholly owned publishing company of Practical Action. Practical Action Publishing trades only in support of its parent charity objectives and any profits are covenanted back to Practical Action (Charity Reg. No. 247257, Group VAT Registration No. 880 9924 76).

The views and opinions in this publication are those of the authors and do not represent those of Practical Action Publishing Ltd or its parent charity Practical Action. Reasonable efforts have been made to publish reliable data and information, but the authors and publisher cannot assume responsibility for the validity of all materials or for the consequences of their use.

Cover photos shows Aerial view of La Vendelée, France
Credit: Francois Boizot, Shutterstock
Typeset by vPrompt eServices, India

Contents

List of figures, photos, tables, and boxes	vii
Acknowledgements	xiii
Foreword	xv

1. Local climate management — 1
   - Introduction — 1
   - Preservation and management — 3
   - Definitions — 5
   - How to use this book — 7

2. Deconstructing and reconstructing the local climate — 9
   - Solar radiation — 9
   - Soil moisture — 14
   - Soil temperature — 18
   - Air temperature — 19
   - Air humidity — 20
   - Wind direction and speed — 21
   - The local climate system — 23

3. Improving the local climate: start of a guiding kit — 27
   - Management of water and moisture — 32
   - Adjusted agronomic practice — 47
   - Management and conservation of vegetation — 60
   - Solar radiation management — 72
   - Wind management — 76
   - Finding the best local climate management practices for a specific site — 80

4. Measuring and monitoring the local climate — 95
   - Local stations — 96
   - Field sensors — 97
   - Remote sensing — 101

5. Conclusion: A patchwork of local action — 107

References — 111
Appendix A: Local climate monitoring tools overview — 133

List of figures, photos, tables, and boxes

Figures

1.1	Three ways to respond to climate change: mitigation (1), adaptation (2), and local climate management (3)	2
1.2	Thermal image of an orchard in Chile with a detail of 10 m × 10 m acquired on 26 December 2021	3
2.1	Overview of the interlinkages of the local climate system and processes relevant for farming practices. This overview is meant as a concept visualization, more detail on the processes is provided in this chapter	10
2.2	Schematic diagram of the long- and shortwave radiation exchanges between a leaf and its environment. Here it can be seen that the sun's radiation (short wave) is received directly and indirectly through reflection by clouds and the soil. In addition, the sky and earth themselves also emit radiation (long wave). Finally, there is also radiation emitted by the plant itself	10
2.3	Albedo effect: comparison of a high albedo surface with a reflection of 80 per cent and a low albedo surface with a reflection of 10 per cent. The lower the albedo the more solar radiation is absorbed instead of reflected into the sky. Only the absorbed radiation can contribute to the heat fluxes	11
2.4	Albedo values for different surfaces	11
2.5	Overview of the surface energy balance. Rn is the net radiation, H is the heat flux that goes into the air, G is the heat flux that goes into the soil, ET is the evapotranspiration	12
2.6	Overview of phase changes from solid, liquid, and gas	13
2.7	Visual representation of diversion of energy fluxes for two different landscapes: desert and highly vegetated area	13
2.8	Visual representation of aspect and slope concepts	15
2.9	Soil water balance of the root zone. The arrows indicate the direction of the water flow. Irrigation water, capillary rise from deeper soil levels, and rainfall contribute positively to this root zone soil water balance. Water is removed from this root zone soil water balance by transpiration (through water uptake by the roots), evaporation, and flow to another soil layer. The soil's water retention and transmission properties control how much incoming water infiltrates and is retained in the soil. If the rainfall intensity exceeds the surface soil's maximum possible	

infiltration rate, ponding water at the soil surface and runoff may occur ... 15

2.10 Schematic overview of soil moisture–temperature coupling. The symbols denote the coupling type: a plus sign (+) stands for positive coupling and a minus sign (–) for negative coupling. The overall coupling between soil moisture and near-surface air temperature is negative (i.e. less soil moisture leads to higher temperatures) ... 17

2.11 Effects of soil temperature on root development. Here it can be seen that the effect of increasing soil temperature stimulates root development until an optimum temperature is reached. At this optimum temperature there is an increased metabolic activity of root cells. Too low soil temperatures reduce the tissue nutrient concentrations which decreases root growth. Too high temperatures can be inhibitive for root development. It should be noted that root development response to temperature is species specific as different species often have different optimum temperatures for root growth ... 19

2.12 Water holding capacity in relation to air temperature. The higher the temperature, the more water vapour the air can hold ... 21

2.13 Wind formation resulting from high- and low-pressure systems ... 22

2.14 Turbulent and laminar flow. When the wind flow is parallel to the ground surface, it is called a laminar flow. When the wind is irregular because of the surface roughness, it is called a turbulent flow ... 23

2.15 Conceptual model of all local climatic components and their interactions relevant for agriculture. In the black boxes, the local climatic components are displayed. In the white and grey boxes, the processes (grey) or factors (white) that influence crop production are shown. The arrows and the signs indicate their relationships: a '+' indicates a positive (increasing) effect, and a '–' indicates a negative (decreasing) effect ... 24

3.1 Visual representation of human–nature interactions between farmers and watershed managers and the local climate ... 28

3.2 Overview of the daytime influence of water conservation practices on the soil temperature, air temperature, air humidity, and soil moisture interactions. A '+' indicates one component having a positive (increasing) effect on the other component and a '–' indicates one component having a negative (decreasing) effect on the other component ... 33

3.3 Functioning of the raised fields ... 46

3.4 An overview of several water and moisture management techniques for local climate improvement implemented in a landscape ... 48

LIST OF FIGURES, PHOTOS, TABLES, AND BOXES   ix

3.5   A thin layer of leaves (left side of the plot) acts as mulch that reduces the loss of soil moisture through evaporation   53
3.6   Top view and cross-section of furrows and ridges   59
3.7   Illustration of seeds planted on the sunny side of the ridge to increase the received solar radiation when the sun's elevation is still low (winter and early spring)   60
3.8   Overview of the hydrological cycle through evaporation and transpiration processes and the biotic pump effect. A '+' indicates one component having a positive (increasing) effect on the other component and a '–' indicates one component having a negative (decreasing) effect on the other component   63
3.9   Examples of various biophysical factors in grassland or cropland (A) and forest (B). Because of cropland's higher reflectivity (albedo), it typically reflects more sunlight than the forest does, cooling surface air temperatures relatively more. In contrast, the forest often evaporates more water and transmits more heat to the atmosphere, cooling it locally compared to the cropland. More water vapour in the atmosphere can lead to a greater number and height of clouds as well as to increased convective rainfall   65
3.10 An overview of several vegetation conservation and management techniques for local climate improvement implemented in a landscape   72
3.11 Visual representation of the effect of the density of a windbreak on the wind flow   79
3.12 Overview of the daytime effect of windbreaks on wind direction and speed and the soil temperature, air temperature, air humidity, and soil moisture interactions. A '+' indicates one component having a positive (increasing) effect on the other components and a '–' indicates one component having a negative (decreasing) effect on the other components   80
4.1   Spatial variation in land surface temperature (LST) in Bole Dugulo, Ethiopia on 10 January 2022. Here it can be seen that a lower tree density (left map) corresponds to a higher LST (right map). The LST is derived from Landsat 8 satellite imagery using the Google Earth Engine script by Ermida et al. (2020)   101

Photos

1.1   The quarry called Latomia del Paradiso in Syracuse, Italy. As a result of a change in the landscape, a 30-metre-deep pit, a local climate that differs from the regional climate is prevailing in the quarry   2
1.2   Mosel Valley, Germany   4

3.1	Road openings interrupt subsurface water flows and increase exposure to wind	29
3.2	Check dams in Ethiopia	34
3.3	Soil bunds along contour	35
3.4	Stone lines combined with zaï pits in Burkina Faso	36
3.5	Eyebrow terraces	37
3.6	Half-moon stone bunds that protect vines from strong winds in La Geria, Lanzarote	38
3.7	Example of water harvesting through semi-circular bunds in Pembamoto, Tanzania	39
3.8	Example of increased vegetation as a result of increased soil moisture levels and soil quality in the semi-circular bund	40
3.9	Trenches	40
3.10	Trenches placed on the contour of a slope adjacent to Lake Ziway, Ethiopia	41
3.11	Agricultural terraces and grazing fields near Foz d'Egua, Piodao, Portugal	42
3.12	Example of a raised field farming system from above	45
3.13	Water storage in abandoned borrow pit in Ethiopia	48
3.14	Example of landscape restoration and water harvesting measures resulting in increased soil moisture accumulation in the rainy season which results in a local climate regulation effect	49
3.15	Example of a pond in the restored landscape. During the day the sun warms the water up. During the night that heat is slowly released and, through dew formation and evaporation, the area around it is cooled and kept moist	50
3.16	Use of straw mulch for crop production	52
3.17	Use of plastic mulch for cucumber and strawberry crops in Urfa, Turkey	53
3.18	Valley blanketed with stones and trees in Harraz, Yemen	54
3.19	Vineyard with a stone terroir in Vinsobres, France: the stones regulate the temperature in the vineyard, insulate the soil and retain heat in the soil, prevent evaporation of moisture from the soil, and enhance dew formation	56
3.20	Intercropping of ginger and maize	57
3.21	Villagers checking on the forest in the evening	64
3.22	Row of metasequoia trees in Makino, Japan. The trees were planted to protect chestnut plantations	66
3.23	Trees grown with the *Kisiki hai* agroforestry technique after three years	68

3.24 Example of an alley cropping system in Hararghe, Ethiopia. Here maize is grown in the alley between chat rows   69
3.25 Grass lines   71
3.26 Shade nets on a small market garden in central Spain   74
3.27 Example of shading in a tree nursery in Amhara region, Ethiopia   75
3.28 Tall green wind shield edges and kiwi cultivation, shot in bright late spring light near Wakamarama, Bay of Plenty, North Island, New Zealand   77
3.29 Aerial view of landscape plots surrounded by shelterbelts serving as windbreak   77
3.30 Hedgerow in front of the cow stable to break down the harsh south-western wind, which creates a more conducive climate for the cows   78
3.31 Farm in Makueni County, Kenya   93
4.1   Example of a weather station on a local site. Weather stations can be equipped with many sensors to conduct a broad range of measurements such as wind speed and direction, air temperature, rainfall, relative humidity, solar radiation, and temperature.   98
4.2   The temperature-moisture sensor placed in the Meshenani Grass Seed Bank   99
4.3   Im Naisreang with her phone, which she uses to irrigate her farm   100
4.4   Example of a drone flying over an agricultural field   104
5.1   Aerial view of the bocage landscape in La Vendelée, Manche, in France. The bocage landscape is characterized by woodland structures, dividing the pastoral land into smaller plots with separated, improved local climates. This photo displays how improving the local climate can be done area by area, consequently covering an entire landscape as a patchwork   108
5.2   Maasai women digging a soil bund for water conservation in the Meshenani Grass Seed Bank as part of the Green Future Farming project   108

Tables
1.1 Classifications of climatological scales   5
3.1 Guidance tool: expected local climatic impacts, the magnitude of impact, and landscape characteristic requirements for each practice   81

Boxes
1.1 The local climate in an ancient quarry   2

1.2	A well-buffered local climate: the Mosel Valley in Germany	4
2.1	Capillary rise: the miracle water buffer	18
3.1	Harmful effects of road development on the Nepalese mountain local climate	29
3.2	Local climate management at the farm	31
3.3	Lanzarote, Spain: wind protection by half-moon-shaped bunds	38
3.4	Lomas de Mejía, Peru: fog collection for ecosystem restoration	43
3.5	The use of water bodies against night frost damage in the Andes	45
3.6	Landscape-wide water harvesting in Tigray, Ethiopia	49
3.7	Harvesting dew in Harraz, Yemen	54
3.8	Stone cover on vineyard in Vinsobres, France	56
3.9	Intercropping to reduce the heat in Preah Vihear, Cambodia	58
3.10	Protecting the rainforest in the Amazon, Ecuador	61
3.11	Madhya Pradesh, India: restoring forest, restoring rainfall	64
3.12	Roadside tree planting	66
3.13	Dodoma, Tanzania: farmer managed natural revegetation	68
3.14	Using shade nets for leafy vegetables in central Spain	74
3.15	Growing citrus in extreme weather conditions in Tajikistan	75
3.16	Silvopasture against weather extremes in Stoutenburg, the Netherlands	78
3.17	Transformed farms in Makueni County, Kenya	93
4.1	Trans-African Hydro-Meteorological Observatory (TAHMO)	97
4.2	Field sensors to detect the microclimatic effects of soil bunds	99
4.3	Remote control farming in Cambodia	100
4.4	Surface Energy Balance Algorithm for Land (SEBAL)	102

Acknowledgements

Giulio Castelli and Lorenzo Villani made significant contributions. Madiha al Junaid designed all infographics.

We acknowledge the insights, wisdom, and contribution of many, but especially Alex Ogelo, Bantamlak Wondmnow, Elena Bresci, Family van Zandbrink, Getachew Engdayehu, Gladys Minoo, Ishan Agarwal, Jackie Kemboi, Justino Piaguaje, Letty Fajardo Vera, Maarten Onneweer, Naluch Lim, Nate Matthews, Pratiti Priyadarshini, Ruth Kimpa, Samwel Jakinda, Tijmen Schults, and Vishakh Rathi.

Many ideas regarding local climate management are developed and tested in the Green Future Farming programme, supported by the Ikea Foundation.

Foreword

Our environmental and social impact in the global age has stretched so far beyond our senses and feelings that we no longer suffer the consequences of our actions as we once did in small close-knit communities living on the bounty of the land. The diffuse and distant impact of our negligent behaviour means we no longer take responsibility for the damage we do to our neighbours and the environment. The restraints of social stigma are now too weak to make us act for the common good.

The 50 gigatonnes of greenhouse gases we emit into the atmosphere each year is the biggest threat to life on our planet and our future wellbeing. The warming effects of the greenhouse are already causing record heatwaves, floods, tornadoes, droughts, bushfires, and climate-change refugees. Millions of farmers in southern Madagascar face crippling starvation. Snow melt on the Himalayas is causing tidal floods, washing away dams, villages, and farms. Searing bushfires in Australia have destroyed thousands of homes and killed millions of animals. Sea level rise in the Pacific is drowning coastal villages and towns on coral islands. Over a quarter of the Great Barrier Reef corals have been bleached by rising sea temperatures, and acidification is compounding the damage to marine life. The elevated temperatures projected by 2050 will cause far greater instability yet to the biosphere which keeps Earth's systems within tolerable limits for life as we know it.

The two main tools we have at hand to combat global warming are mitigation – cutting emissions by switching from fossil to renewable energy – and adaptation, redesigning and relocating our homes, towns, infrastructure, and food production and water supply systems to adjust to climate-induced changes. Despite the agreements reached at COP26, the pledges fall far short of the massive switch from fossil fuels to renewable energy needed to keep global temperatures within a tolerable rise of 1.5°C by 2050. Governments are too wary of the economic disruption caused by a rapid transition, and big industry too invested in fossil fuels and shareholder dividends to take decisive action.

I have long argued that climate warming, though the biggest threat to the future of our planet, is dwarfed so far by the far greater damage we have done already to the biosphere. The damage due to overuse, large-scale agro-industrial farming, and industry to habitat, biodiversity, rivers, and lakes, and through soil erosion and nutrient leaching, now far exceeds all natural forces governing the planet. So enormous is our footprint that earth systems scientist, Paul Crutzen, dubbed our transformation of the Earth a new epoch, the Anthropocene.

Ecological degradation, the loss of the land's ability to decompose and recycle nutrients and replenish itself after hard years, diminishes both the wellbeing of humanity and the natural capacity of the Earth to buffer the lithosphere, atmosphere, and biosphere from climatic extremes.

Great as our degradation of the land is, we can do something about it in a matter of years rather than decades. Restoring land health is a win-win for the farmer, rancher, the economy, society, and the biosphere. It combines both mitigation and adaptation measures and opens up a third avenue by scaling up local individual actions to a global movement to combat climate warming. Restoration involves no massive investment or sophisticated technology, and the returns are quick and visible.

Whether in our backyard or schoolyard, field, estate, or in our forests, wetlands, and rangelands, we can each manage our patch of land. The restoration efforts of billions of people around the world can replenish the good earth and give us pride in doing our bit.

The scope for restoration is enormous across the half-earth we have damaged through bad land practices. The basic principles of sustaining and restoring the land are deeply embedded in societies which have learned to live within the limits of their environment. The husbandry practices of Hanunóo cultivators of the Philippines, Konso farmers of the Ethiopian Highlands, and pastoralists of East Africa show that farmlands and ranchlands can be sustained for millennia where communities have strong cultural governance practices and ecological know-how.

Good land stewardship creates positive feedback cycles between plant cover, soil organic matter, and water retention, resulting in less erosion and a cooler, more humid local climate. Restoring local land health and local climates also boosts farm and ranch production.

This is the message of *Managing the Local Climate*. Here is a how-to manual that compiles a catalogue of methods for restoring land health and, in aggregate, combating global warming. The methods and lessons of successful land restoration, once confined within local communities, can now be shared around the world, and must be. Social media has opened a new way of connecting people within, between, and across communities to learn from local success that can grow to a global movement. This is the message of Nobel Laureate Wangari Maathai's Green Belt Movement.

Dr David (Jonah) Western
Founding Executive Director of the African Conservation Centre

CHAPTER 1
Local climate management

1.1 Introduction

This book discusses the management of local climates, a topic that – surprisingly – is often forgotten in the discussion on global climate change despite the huge rebalancing opportunities. Global climate change, triggered by changes in the composition of the world's atmosphere and the strength of solar radiation cycles, is a significant concern of the present time. The two common responses to global climate change are to *mitigate* and reduce the emission of greenhouse gases or to *adapt* to a world with higher temperatures, more frequent flooding, and more intense droughts (IPCC, 2021). Both these responses are, in a sense, defensive: they either reduce the problem or find the best way to live with it.

Yet this book describes an additional third way to respond to climate change: the preservation and better management of local climates. This approach distinguishes itself by being bottom-up and proactive. It influences climate at a local level, in contrast to mitigation strategies operating to reduce the warming of the global climate. Local climate management is also proactive and not passive. It does not merely react to changed climate circumstances as in adaptation but empowers people to create their own local climates that are more amenable for local functions and better harnessed against climate change extremes. Hence, managing local climates is a third way, complementing mitigation and adaptation (Figure 1.1).

Local climates are in several ways what matter most: they are what we experience in reality. It is where humans live and plants grow, where biodiversity and ecology happen. Here a dynamic interaction of forces determines the moisture available to the different ecosystems, the local rainfall, the presence of dew and frost, the temperatures for plant growth and germination, the vigour of soil biotic life, capacity to fixate nitrogen by soil biota, and the occurrence of pests and diseases. Local climates are where meteorology lands on earth (Ismangil et al., 2016). Local climates thus alter from the prevailing regional climate and are the result of heterogeneity in landscapes; Box 1.1 shows an example.

2 MANAGING THE LOCAL CLIMATE

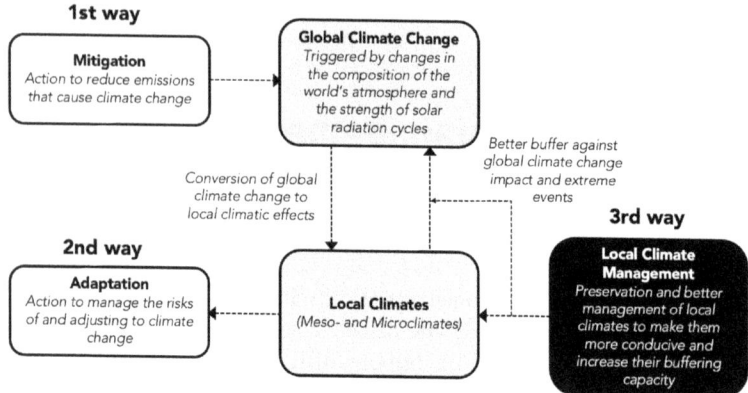

Figure 1.1 Three ways to respond to climate change: mitigation (1), adaptation (2), and local climate management (3)

Box 1.1 The local climate in an ancient quarry

Landscape morphology creates unique local climates, as visible in abandoned quarries. The quarries built in the town of Syracuse in Sicily during the ancient Greek period have a long and painful history for the enslaved people and prisoners deployed to chisel away 30-metre layers of limestone under often atrocious circumstances.

The quarry in Neapolis Park now presents a unique climate – cooler, wetter, and protected from desiccating winds. In a reversal of its painful past, it is now called *Latomia del Paradiso* – Paradise Pit.

Photo 1.1 The quarry called Latomia del Paradiso in Syracuse, Italy. As a result of a change in the landscape, a 30-metre-deep pit, a local climate that differs from the regional climate is prevailing in the quarry

1.2 Preservation and management

What is very important is that local climates are not the passive outcomes of global weather phenomena but that they can be managed by a range of farm operations, land, water, and regreening measures, and nature-based solutions. These measures buffer against larger climate change trends and absorb the disruptions, and their cumulative impact can partly counterbalance the predicted 1.5–2.5°C temperature rise in the coming decades. Research conducted in Ethiopia, for instance, established that local landscape restoration reduced local land surface temperatures by almost 2°C (Castelli et al., 2018). A study on farmer-implemented revegetation measures in dryland Tanzania showed a decrease in land surface temperatures of close to 1.5°C (Villani et al., 2020). Thermal radiometers can measure the spatial variability of land surface temperature. The example in Figure 1.2, from a midsummer day in Chile, indicates the variability of more than 5°C due to different on-farm practices in an irrigated vineyard. This temperature buffering effect shows that local climate management can neutralize the impact of global climate change.

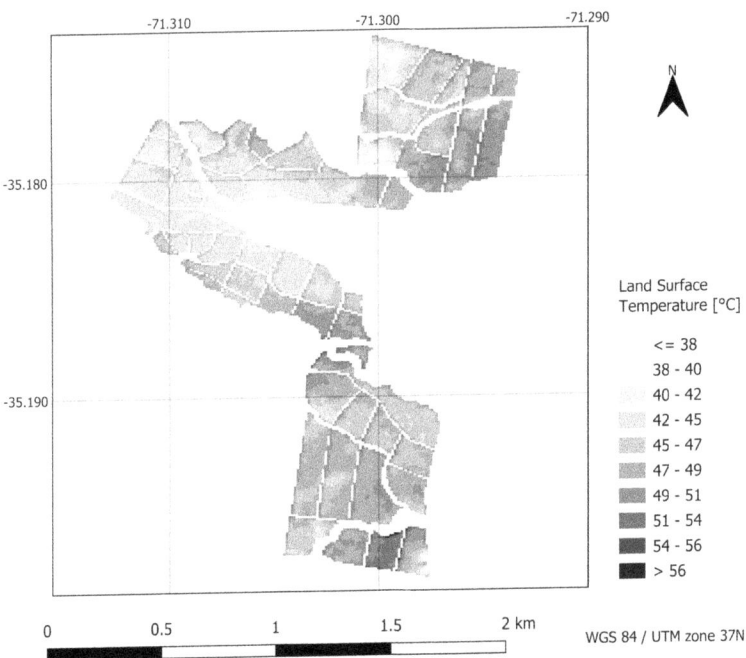

Figure 1.2 Thermal image of an orchard in Chile with a detail of 10 m x 10 m acquired on 26 December 2021
Source: Map created by Tijmen Schults with data provided by Irriwatch

Moreover, good local climate management creates more conducive conditions for agricultural production and other functions (Kingra and Kaur, 2017; Hadid and Toknok, 2020). Increased agricultural productivity results from, for example, more secure moisture levels, conducive wind conditions, shelter from temperature extremes, or soil temperatures that stimulate biotic life and promote root growth. Local climate conditions, in particular air humidity, also have a large bearing on the incidence of local pests.

There is a wide range of measures available to manage the local climate. Still, they all start with appreciating and preserving what is there; that is, those landscape elements that contribute to stable and beneficial local climate – particularly local vegetation, forests, and water bodies. In addition, there is a range of other measures that can be applied, as explained later in this book. The range of options differs from area to area. Sites with a sheltered topography, unique geology, adequate water resources, and greenery will be better buffered. Box 1.2 is an example. On the other hand, in dry, barren, and flat savannah areas, the option to create conducive, well-buffered local climates is more challenging, and climate variation may be amplified in temperature peaks, drought streaks, or episodic rain and flooding events. However, as this book shows, opportunities to better manage the local climate occur almost everywhere.

> **Box 1.2** A well-buffered local climate: the Mosel Valley in Germany
>
> For at least 2000 years, the Mosel Valley in Germany has been one of Europe's northernmost wine-growing regions, famous for its Riesling and Elbling varieties. The sheltered mountains and the presence of the winding Mosel River create a mild local climate buffered against temperature extremes, which is very suitable for viniculture. Moreover, the slate geology causes the daytime heat to be absorbed only to be gradually released at night, creating even more conducive local climate conditions that enable viniculture at this latitude.
>
>
>
> **Photo 1.2** Mosel Valley, Germany

Another essential argument for taking local climate management seriously is that there are many techniques to monitor and map the different constituent components that make up the local climate at the farm level and landscape scale. In other words, we have the means to measure local climate in great detail and develop a systematic approach. Many of these monitoring techniques – high-resolution remote sensing and new generation field sensors – can be used immediately.

1.3 Definitions

Climates can be categorized into the macroclimate, the mesoclimate, and the microclimate (Table 1.1). In practice, there is some overlap between the three different climate scales. This book defines the meso- and microclimate as the local climate which can be influenced by the land surface, meaning that changes in a landscape directly influence the climate of that area.

The macroclimate

The macroclimate is the climate of an area or zone characterized by long-term observations. Features that influence the macroclimate are, for example, the distance from the sea, the seasons, and the latitude and altitude. These are characteristics that cannot be altered but still influence the mean of meteorological variables over time. For example, in general it is found that areas with a higher latitude and altitude tend to have cooler temperatures. Or areas close to the sea tend to have more moderate climates than areas further from the sea. These macroclimatic features also result in several global-scale climatic phenomena such as El Niño and La Niña, which are the warm and cool phases of a recurring climate pattern across the tropical Pacific.

Table 1.1 Classifications of climatological scales

Climatological scale		Vertical scale	Horizontal scale
Global climate	Macroclimate	Free atmosphere	100 to 10,000 kilometres (Hupfer, 1989)
Local climate	Mesoclimate	Atmospheric boundary layer *(Varies between 100 metres and 3 kilometres (Stull, 1988))*	<100 kilometres (Hupfer, 1989)
	Microclimate	Atmospheric surface layer *(lowest 10% of the atmospheric boundary layer (Moene and Van Dam, 2014))*	<100 metres (Hupfer, 1989)

Global climate change, the result of an increased concentration of carbon dioxide in our atmosphere that is amplifying Earth's natural greenhouse effect, is increasing overall temperatures and altering seasonal patterns. These macroclimatic changes may be ameliorated or exacerbated by local climates, making local climate management an important approach in counterbalancing the macroclimatic temperature rise and the accompanying meteorological changes in the coming decades.

The local climate

Unlike the macroclimate, the local climate is directly influenced by the land surface. On a vertical scale, this is determined by the atmospheric boundary layer. This atmospheric boundary layer is the lower part of the atmosphere where the nature and properties of the surface affect turbulence directly (Brutsaert, 1982). Here local climates result from the fine-grained, localized, dynamic interplays between different surface layer processes, particularly those by which net radiation is converted into local heat and water vapour fluxes (Foken, 2008).

Local climates consist of multiple components: solar radiation, soil moisture, soil temperature, air temperature, air humidity, local rainfall, and wind direction and velocity. All components are inextricably linked; changing one factor will also affect the other parts. There are two climatological scales within the local climate below this atmospheric boundary layer: the mesoclimate and the microclimate.

The mesoclimate

The largest scale within the local climate is the mesoclimate. Albeit to a lesser extent than the microclimate, the mesoclimate is still influenced by the landscape surface and is determined by the atmospheric boundary layer. The mesoclimate ranges at a horizontal scale up to 100 kilometres (Hupfer, 1989). Multiple microclimates are boxed together to make up the mesoclimate. In reality, these spatial scales overlap, and the various 'climates' integrate and are interdependent (Sharpe, 1987).

The microclimate

The lowest 10 per cent of the atmospheric boundary layer is the surface layer. This layer is characterized by large differences in temperature and wind speed resulting from the heterogeneous land surface. The atmospheric surface layer determines the vertical scale of microclimates (Moene and Van Dam, 2014). Microclimates have a horizontal spatial scale up to 100 metres (Hupfer, 1989), spanning the spatial scale from single leaves up to hillslopes (Sharpe, 1987). The microclimate is the climatic scale most strongly associated with processes occurring at the surface layer: for example, energy and matter exchange, and radiation processes close to the ground surface (Foken, 2008).

The local variability in earth-atmosphere interfaces thus results in a patchwork of different microclimates. Even in a small garden, a microclimate may persist; hedges may act as a windbreak and reduce evaporation resulting in a wetter and cooler site, and rocks in the garden may get warmed up by the sun during the day and serve as a heater during the night. Modifying the land use and soil practices and implementing practical interventions such as windbreaks and shade nets thus directly alter the microclimate.

1.4 How to use this book

Local climates – like many other micro-level phenomena – often go unnoticed. Promising research has been done on different elements (De Frenne et al., 2013, Kingra and Kaur, 2017, Villani et al., 2020), but the total picture may be elusive. There is underutilized potential in practising systematic and interdisciplinary local climate management to buffer global climate change. We may want to look at the world not as the passive recipient of global climate but as a patchwork of local climates that can be managed one by one. As Judith D. Schwartz asks in her book *The Reindeer Chronicles*, 'How many microclimates does it take to make a new climate?' (Schwartz, 2020: 61).

This book is meant as a guide on how to intentionally sustain and create conducive and well-buffered local climates by preservation and by proactive, well-considered, and measurable action. We hope the book contributes to a systematic approach to managing the local climate in landscapes, akin to addressing urban heat islands (Brown, 2011).

Whereas the local climate interconnects with many ecological services, the bias in this book is towards agriculture, the largest human-made land use. The book discusses how to create conducive local climate conditions for farming with local measures, but also how to use local agriculture – cropland farming, agroforestry, and agricultural water storage – as an instrument in building up better-buffered climates. This is not to say that the local climate is not paramount for other ecological services, such as biodiversity or drought resilience; it should also be managed as such. We hope that practical approaches for all these services will become more and more elaborate over time.

This book discusses the main mechanisms that shape local climate components, the practices that influence it, and the measuring methods.

The next chapter (Chapter 2) deconstructs the local climate into six constituent parts and discusses the relevant micrometeorological processes. Understanding local climate components and their mutual relations provides the building blocks for understanding how different land and water management practices transform microclimates and mesoclimates.

Chapter 3 is the beginning of a guiding kit for local climate management. This chapter discusses what can be done and elaborates on how land, water management, and vegetation interventions and practices influence the local climate system and, as a result, agroecological conditions. Hopefully, interventions in this field will expand in range and appropriateness over time. This chapter also provides examples of local climate management, as practised in different parts of the world, showing that local climate management is possible and effective. The cases show the systems' often intricate working and the use of local climate to create more conducive agroecological systems.

Chapter 4 elaborates on monitoring and mapping the local climate. This chapter demonstrates how remote sensing, climate data sets, local weather stations, and field sensors can support a systematic and careful introduction of positive measures. Ultimately, this book aims to move climate change concerns towards the local scale by strengthening micro and mesoclimate management and improving agroecological conditions. The aim is to create a global mosaic of resilient and beneficial local climates.

CHAPTER 2
Deconstructing and reconstructing the local climate

To effectively improve the local climate requires a better understanding of the main elements of the local climate. In this chapter, we break down the local climate into six main physical processes and principles: solar radiation, soil moisture, soil temperature, air temperature, air humidity, and wind speed and direction (Figure 2.1). These are the building blocks for understanding how the local climate works and how local practices can positively contribute, as will be discussed in Chapter 3. The last section of this chapter reconstructs the interaction between the different building blocks in local climate management.

2.1 Solar radiation

Local ecosystems are open thermodynamic systems in which energy flows in and out. In the form of electromagnetic radiation, this energy consists of a stream or flow of particles. Each particle has an energy content – the greater the frequency (shorter wavelength), the greater the particle's energy content. The sun is the primary source of all energy that may be considered primarily short-wave radiation (Rosenberg et al., 1983).

Many atmospheric factors, such as the cloud cover, aerosol presence, and water vapour content, affect how much solar radiation the surface receives (Shen et al., 2008). Under clear sky conditions, water vapour and aerosols are the main factors resulting in the loss of total solar radiation (Zhang and Wen, 2014). The earth itself also releases radiation, which is called long-wave radiation. Part of this radiation is returned to the Earth's surface as it gets reflected by gases in the atmosphere consisting primarily of water vapour and carbon dioxide (Figure 2.2); this is also the primary process driving climate change.

As shown in Figure 2.2, some of the incoming solar radiation gets reflected into the sky, and this proportion is called the albedo. Johann Heinrich Lambert first introduced the term albedo in his 1760 work *Photometria*. The darker a surface, the lower its albedo and

10 MANAGING THE LOCAL CLIMATE

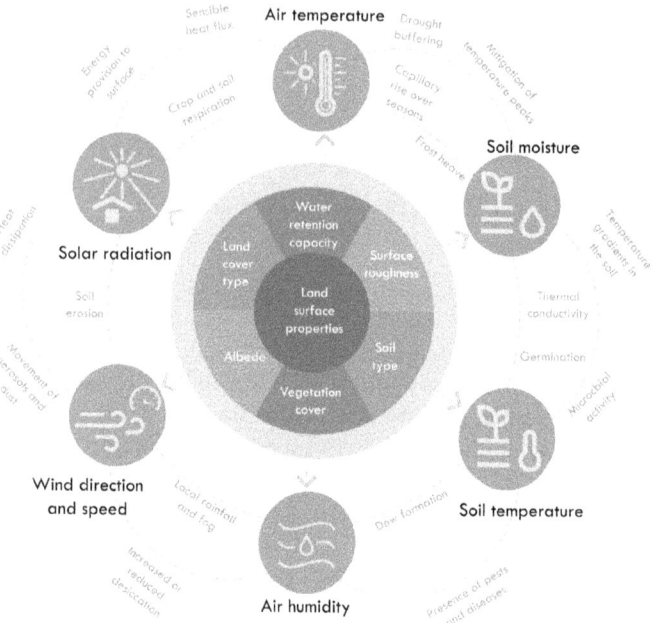

Figure 2.1 Overview of the interlinkages of the local climate system and processes relevant for farming practices. This overview is meant as a concept visualization, more detail on the processes is provided in this chapter
Source: MetaMeta

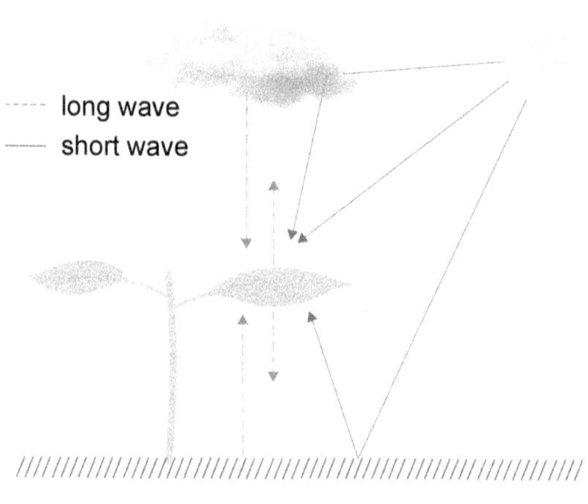

Figure 2.2 Schematic diagram of the long- and shortwave radiation exchanges between a leaf and its environment. Here it can be seen that the sun's radiation (short wave) is received directly and indirectly through reflection by clouds and the soil. In addition, the sky and earth themselves also emit radiation (long wave). Finally, there is also radiation emitted by the plant itself

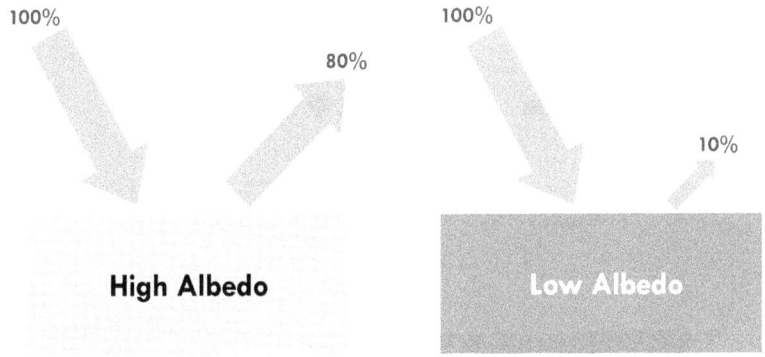

Figure 2.3 Albedo effect: comparison of a high albedo surface with a reflection of 80 per cent and a low albedo surface with a reflection of 10 per cent. The lower the albedo the more solar radiation is absorbed instead of reflected into the sky. Only the absorbed radiation can contribute to the heat fluxes

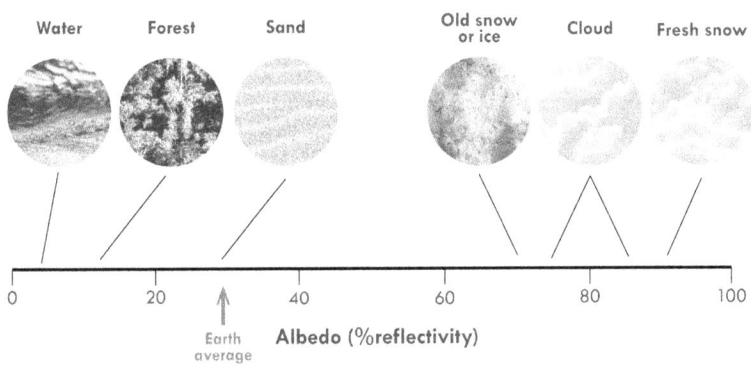

Figure 2.4 Albedo values for different surfaces
Source: Adapted from Earle, 2019

thus its reflection, meaning that more solar radiation is absorbed and contributes to heating processes (Lambert, 1760) (Figure 2.3).

The albedo can vary from less than 10 per cent for a coniferous forest to 95 per cent for fresh snow. Generally, most crop fields have an albedo of around 25 per cent (Unwin and Corbet, 1991). Water surfaces are poor reflectors (low albedo) and thus serve as a good sink for solar energy (Rosenberg et al., 1983) (Figure 2.4).

Thus, when a surface is not entirely white, some solar energy is absorbed and contributes to the energy balance. This part of the solar energy is called net radiation. Net radiation drives all local climatic processes such as air temperature differences, winds, soil

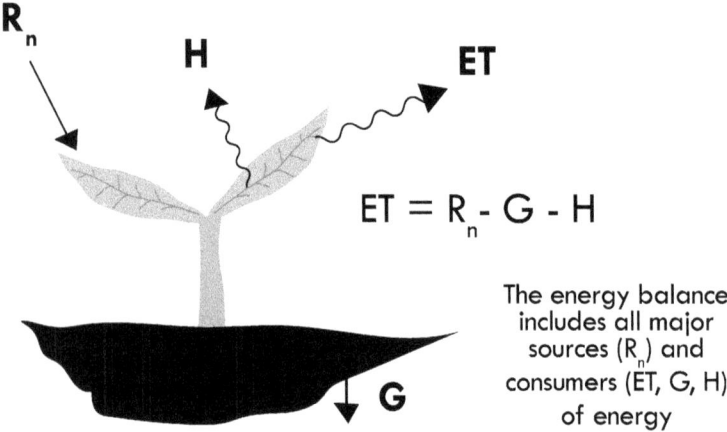

Figure 2.5 Overview of the surface energy balance. Rn is the net radiation, H is the heat flux that goes into the air, G is the heat flux that goes into the soil, ET is the evapotranspiration
Source: Wim Bastiaanssen

temperature, evapotranspiration, and rainfall (Rosenberg et al., 1983). The conversion of this energy into the heating of air and soil creates the local climatic systems. These systems are intricate, but their structural elements function based upon a relatively small set of fundamental principles.

An overview of this surface energy balance is provided in Figure 2.5. The source of energy in the system is net radiation (R_n). The consumers of the energy are the processes of heating the air (H), heating the soil (G), and evapotranspiration (ET).

The energy flux that contributes to the heating of air and soil (H and G) is called the sensible heat flux. Sensible heat is the flux related to changes in the temperature of a gas or object with no change in phase. Evapotranspiration (ET), on the other hand, is called latent heat flux. Latent heat is related to changes in phase between liquids, gases, and solids. When water vapour condenses and becomes liquid, and when water freezes, energy is released from the water to the surroundings/atmosphere. When frozen water melts and the liquid evaporates, energy is absorbed from the surroundings/atmosphere (Figure 2.6). Evaporation thus cools the environment by requiring energy from the surroundings/atmosphere (Unwin and Corbet, 1991).

All net radiation must go somewhere: it either contributes to evapotranspiration processes (latent heat flux), or it contributes to the heating of the air and soil (sensible heat flux). This ratio of energy

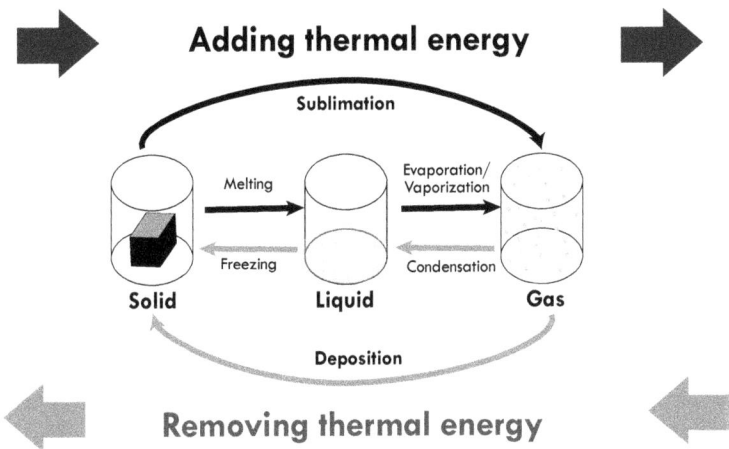

Figure 2.6 Overview of phase changes from solid, liquid, and gas

fluxes from one state to another by sensible and latent heat is called the Bowen ratio (Bowen, 1926).

The horizontal distribution of Bowen ratios depends on surface heterogeneity (Friedrich et al., 2000). This means that how the net radiation is diverted into these two energy fluxes depends on landscape characteristics such as albedo, vegetation cover, and soil moisture availability. Figure 2.7 displays this phenomenon for two different situations: a desert landscape with low soil moisture and little vegetation and a highly vegetated landscape with sufficient soil moisture. There is little evapotranspiration for the desert landscape as

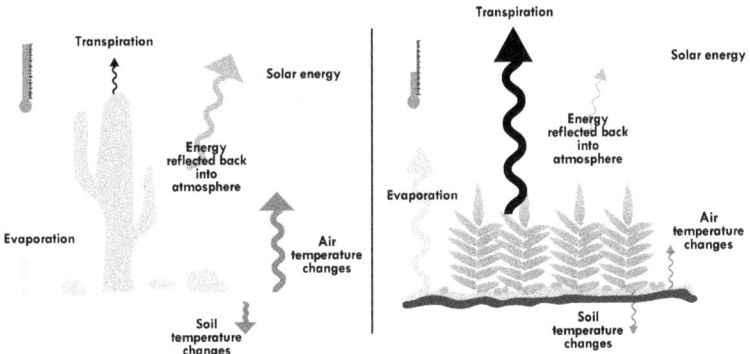

Figure 2.7 Visual representation of diversion of energy fluxes for two different landscapes: desert and highly vegetated area
Source: Wim Bastiaanssen

the moisture at the surface is limiting, meaning that most net radiation is used to heat the surface and air. For the vegetated area, the evapotranspiration fluxes are more prominent as the moisture levels at the surface are not limiting. Only a little of the net radiation contributes to the heating of air and soil. This example shows how vital the outlook of a landscape is in determining the partitioning of net radiation: will it contribute to heating the soil and air? Or is there enough soil moisture and vegetation to enable evapotranspiration processes resulting in cooler temperatures?

Large quantities of energy from the sun contribute to the energy balance. By day, the incoming radiation generally exceeds outgoing radiation. The surplus of this radiation energy flows into the ground and air and contributes to these processes of evapotranspiration and condensation. When incoming radiation is reduced by night, the situation is generally reversed, and energy is lost from the surface by outgoing radiation (Geiger et al., 2003).

The quantity of solar radiation that reaches a horizontal unit of the Earth's surface and contributes to this energy balance depends on several factors: the intensity of radiation emitted by the sun, the sun's position compared to the Earth, and the transparency of the atmosphere (Rosenberg et al., 1983). The sun's position compared to the Earth can be expressed with Lambert's cosine law. This law states that when the sun is at its highest point, radiation is at maximum, and the cosine of the sun's angle to a line at right angles to the surface is equal to one. When the sun is at the horizon, the cosine is equal to zero and the radiation received from the sun is approximately zero (Lambert, 1760). In the landscape, two factors influence Lambert's cosine law and determine the flux density at the surface: aspect (compass direction) and inclination or slope (steepness of angle with the horizontal plane) (Figure 2.8).

Solar radiation, particularly the net radiation that contributes to the energy balance, is thus of high importance in determining the local climate, with the albedo, aspect, and slope of a plot being significant factors affecting the amount of net radiation in the energy balance.

2.2 Soil moisture

As the example in the previous section (Figure 2.7) showed, the availability of soil moisture in a plot is essential for determining the local climate of that plot. Soil moisture is the amount of water per cubic metre of soil. The water content in any soil layer can decrease by soil evaporation, root absorption, runoff, outgoing subsurface flow, or flow to an adjacent soil layer (Figure 2.9).

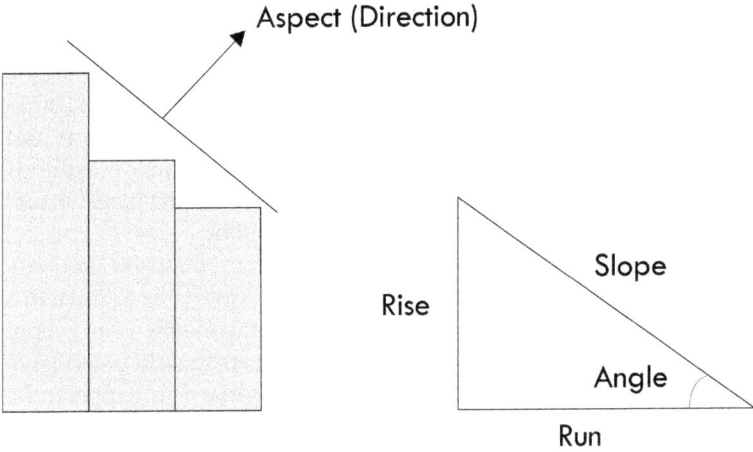

Figure 2.8 Visual representation of aspect and slope concepts

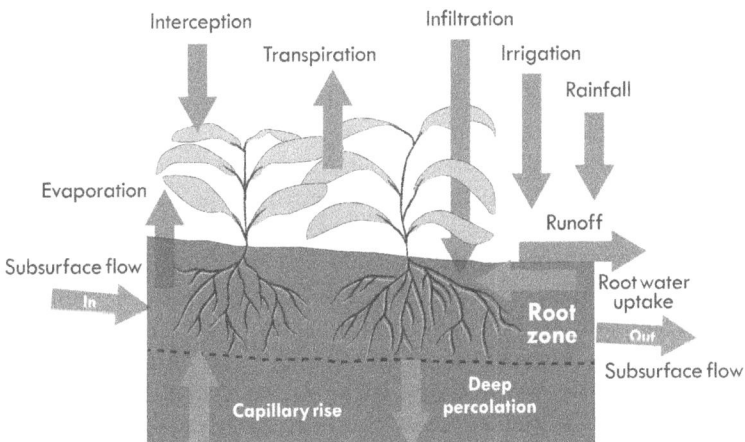

Figure 2.9 Soil water balance of the root zone. The arrows indicate the direction of the water flow. Irrigation water, capillary rise from deeper soil levels, and rainfall contribute positively to this root zone soil water balance. Water is removed from this root zone soil water balance by transpiration (through water uptake by the roots), evaporation, and flow to another soil layer. The soil's water retention and transmission properties control how much incoming water infiltrates and is retained in the soil. If the rainfall intensity exceeds the surface soil's maximum possible infiltration rate, ponding water at the soil surface and runoff may occur
Source: Adapted from Allen et al. (1998)

Schwingshackl et al. (2017) provide an overview of studies on the role of soil moisture in climate variability. Various studies showed that soil moisture impacts the evolution of near-surface air temperature, the formation of precipitation, and the carbon cycle. Soil moisture can substantially affect the appearance and severity of droughts and heatwaves. This importance of soil moisture for atmospheric conditions arises from its control on water and energy fluxes at the land surface, altering the atmosphere's water and energy content.

When the soil moisture level decreases, the evapotranspiration rate is reduced below the potential since less water is available to drive this process. Less energy goes into the latent heat fluxes when the evapotranspiration rates are reduced. As a result, more energy will be available for the sensible heat flux, increasing the near-surface air temperature. When the soil is wet, evapotranspiration is possible, which consumes the incoming energy. This increase in latent heat flux decreases the sensible heat fluxes which results in lower near-surface temperatures (Fritschen and Van Bavel, 1962; Bouchet, 1963) (Figure 2.10).

The latent heat flux consists of both evaporation and transpiration. Crop transpiration is a central process in plant production, and it directly influences agricultural yields (Smith and Steduto, 2012). From a plant physiology viewpoint, soil and open-water evaporations, on the other hand, are wasteful processes.

Soil moisture availability thus determines the partition of energy between sensible, latent, and soil heat fluxes. In turn, these heat fluxes determine the temperature and humidity of the local climate (Hutjes, 1996). Areas with available soil moisture have a more balanced local climate where air and soil temperatures are buffered against climatic extremes. Aerts et al. (2004) suggested using this thermal buffer capacity of an ecosystem as an indicator of the restoration status of protected areas in the Northern Ethiopian Highlands. The better an ecosystem can dissipate solar radiation energy flows, the higher the restoration stage of that ecosystem. A more balanced local climate facilitates plant growth and affects the weather and local rainfall patterns.

Soil moisture in itself is also essential for soil microbial activity (Manzoni et al., 2012; Meisner et al., 2015). Microbial activity diminishes at low soil moisture levels across biomes and soil types. Likewise, atmospheric nitrogen fixation is reduced when soil moisture levels drop (Zullu Jr et al., 2008). Optimal conditions for micro-organisms to break down organic matter and release nutrients are when moisture takes up around 60 per cent of the open water pore space. Too much soil moisture prevents oxygen supply, leading to slower microbial activity (Ismangil et al., 2016).

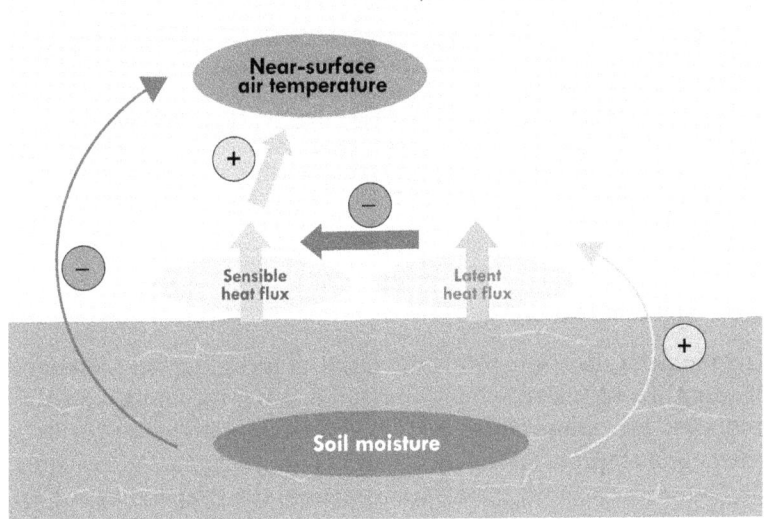

Figure 2.10 Schematic overview of soil moisture–temperature coupling. The symbols denote the coupling type: a plus sign (+) stands for positive coupling and a minus sign (−) for negative coupling. The overall coupling between soil moisture and near-surface air temperature is negative (i.e. less soil moisture leads to higher temperatures)
Source: Adapted from Schwingshackl et al. (2017) © American Meteorological Society. Used with permission

Another effect of sufficient soil moisture levels is reduced soil erosion by wind. The bond between soil particles becomes stronger through capillary and adhesive forces of absorbed water molecules surrounding the soil particles. Wind erosion can lead to soil removal around the shallow parts of the crop roots system, also called uprooting, causing instability and water and nutrient uptake problems (Jones, 2014). Yield reductions of 9 to 18 per cent for maize and 17 to 24 per cent for soybean on three soils subject to severe erosion have been reported. This yield reduction was mainly caused by reducing soil depth and soil water holding capacity (Gardner et al., 1999).

Increased soil moisture beyond field capacity may also increase flow to the adjacent soil layers and enable shallow groundwater recharge. When the deeper soil moisture levels are replenished after a rainy period, soil pores transport water back upward because of the capillary action within the soil. The recharged shallow groundwater then becomes available for plant root uptake. Specific circumstances may trigger this capillary rise to benefit crop production outside the rainy season (Box 2.1).

> **Box 2.1** Capillary rise: the miracle water buffer
>
> In Sudan's Gash Delta Spate Irrigation Scheme, farmers can cultivate watermelon as a second crop at the end of October – long after the rains have fallen, and the floods have been diverted. In the drylands of Sindh in Pakistan, the same can be observed: chickpeas are grown as a second crop on seemingly zero rainfall many months after the spate floods have been diverted to irrigate the land (Kool et al., 2016). In the cool dry season, due to changed atmospheric pressure, soil moisture moves up by capillary rise from lower layers wetted in the rain and flood season 3–5 months earlier.

2.3 Soil temperature

The temperature of the soil is the result of incoming net radiation. Shading or a high albedo rate will result in lower incoming radiation and thus less temperature rise. A vegetation canopy serves as a reflector or absorber for the incoming radiant energy before reaching the soil. Because of the plant's transpiration, the temperature rise on a vegetation canopy will be lower than that of bare soil of about the same reflectivity (Stoutjesdijk and Barkman, 1992). Vegetation will thus hinder the increase in ground temperature in the daytime and cooling at night, resulting in fewer temperature extremes (Scott, 2000).

Temperature conditions within the soil are continually varying. The heat exchange processes primarily at the soil surface are responsible for this soil temperature fluctuation. Heat constantly moves into or out of the soil, and thermal energy is redistributed in the soil (Rosenberg et al., 1983). The system attempts to reach equilibrium, but heat inputs and heat sinks continually perturb it. The ratio of this soil temperature variation is highly dependent on the thermal conductivity of the soil. Thermal conductivity is a measure of the soil's ability to conduct heat. Water has high thermal conductivity, and is 20 times more conductive than air (Bonan, 2016). Whether soil pore space is filled with air or water is significant in determining thermal conductivity. In light soil, such as sandy soil, containing much air, the surface temperature increases rapidly in the daytime and decreases quickly at night. Soil temperature extremes will thus be modified when the moisture content is sufficient as the water serves as a heat sink (Van Wijk, 1963; Yoshino, 1975). This buffering capacity protects the plant root system against sharp and sudden changes in temperature. The effect of soil temperature on root development is displayed in Figure 2.11.

A buffered soil temperature is essential for many other processes. Too high soil temperatures can stall the biological processes of

Figure 2.11 Effects of soil temperature on root development. Here it can be seen that the effect of increasing soil temperature stimulates root development until an optimum temperature is reached. At this optimum temperature there is an increased metabolic activity of root cells. Too low soil temperatures reduce the tissue nutrient concentrations which decreases root growth. Too high temperatures can be inhibitive for root development. It should be noted that root development response to temperature is species specific as different species often have different optimum temperatures for root growth

micro-organisms, even kill them. On the other hand, extremely low temperatures also influence the soil microbial population, the rate of organic matter decomposition, inhibit water and nutrient uptake by plants, inhibit nitrification, reduce soil fertility, increase desiccation when air temperatures are higher, and diminish the role of larger fauna in soil structural development (Gliessman, 2015). This highlights the importance of buffering soil temperatures with local climate management.

2.4 Air temperature

Local air temperature is a result of incoming and outgoing net radiation. A considerable amount of radiation energy is required to heat the soil and plants from sunrise. Sensible heat fluxes increase the air temperature after these surfaces become warm relative to the air above. The sensible heat flux is usually away from the surface during the day and towards the surface during the night; then, the air is generally warmer than the surface (Rosenberg et al., 1983).

Several on-site modifications affect the air temperature. First, as we have seen in the previous sections, vegetation can increase transpiration and lower the local air temperature. Second, moisture in the soil can decrease the overall air temperature and mitigate extreme temperature, as was explained earlier (Bouchet, 1963). Third, the reflectivity of the soil, the albedo, is another critical factor in determining the local air temperature since this determines how much radiation is absorbed and thus contributes to the heating of the soil and, consequently, the air. The lighter the colour of the soil, the more solar radiation will be reflected and the lower the soil and air temperature.

2.5 Air humidity

Air humidity is the amount of water vapour that results from evapotranspiration induced by net radiation and soil moisture. It is the basis for cloud, fog, and dew formation, making air humidity an essential component of the local climate.

There are several ways to express air humidity. The most used parameter is relative humidity, usually expressed as a percentage. Relative humidity refers to the amount of water vapour in the air compared to what it can potentially hold. Thus, relative humidity of 25 per cent indicates that the atmosphere holds 25 per cent of the total water vapour it could hold at full saturation. Relative humidity of 100 per cent means that the air is saturated, and mist, fog, and clouds will form. The air temperature determines the amount of water vapour that air can hold, with warm air holding more water vapour than cold air (Figure 2.12).

Air humidity plays a significant role in evapotranspiration processes. An accurate way to express the driving force of water loss from plants and soil is the vapour pressure deficit (VPD). This is the difference between the amount of moisture in the air and the amount of moisture that the air could hold when it is saturated. This pressure drives plant transpiration and soil evaporation. A high VPD means that the air can still hold a large amount of water, enabling the plants to transpire. When the VPD is low, the air is near saturation, and plants cannot transpire effectively (Jones, 2014). However, a high VPD can also result in increased water loss rates from moist soils, causing drying and heating of terrestrial surfaces and contributing to more frequent and severe drought events and plant water stress (Dai, 2013). Too high VPD can also cause plants to close their stomata to minimize water loss, resulting in reduced photosynthesis (Franks et al., 1997).

As stated earlier, air humidity is the basis for cloud, fog, and dew formation. Condensation is the water-cooling process whereby water changes from the gas state to the liquid form. Dew is formed when air temperature with a specific moisture level cools below a certain temperature. The air then becomes too cold to carry all this moisture and dew will form. This temperature is called the dew point temperature. An example can be found in open savannah areas that

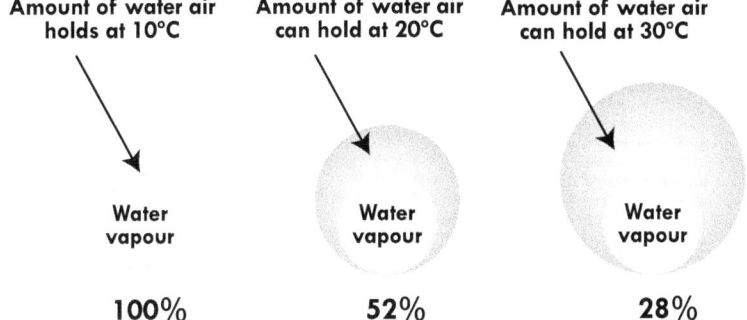

Figure 2.12 Water holding capacity in relation to air temperature. The higher the temperature, the more water vapour the air can hold

provide favourable conditions for dew formation as air quickly cools down in the case of low wind velocities and long-wave up-welling radiation. Here the dew formation in the morning is an essential source of water for ground vegetation and animals (Foken, 2008).

Fog can also be an essential moisture source for plant growth in semi-arid environments. Fog forms when the atmospheric water vapour concentration reaches the saturation point (Rosenberg et al., 1983). Fog formation happens regardless of conditions at the surface, making fog a purely atmospheric process.

It should be noted that most air humidity, especially in arid regions, is formed from moisture evaporated from the immediate ground surface. Under these conditions, air humidity is essentially recovered soil moisture and does not represent a new addition of water to the local system (Wilken, 1972). However, this is water that would otherwise have been lost from the system, which makes recovering this water still a significant contribution.

Air humidity also plays a vital role in the formation of precipitation. When the relative humidity reaches saturation, clouds will form because water vapour gas condenses into visible water droplets. When these droplets grow too large, they fall and result in rain (Wilken, 1972).

2.6 Wind direction and speed

Winds are the results of differences in pressure fields. Air movement is generated through pressure gradients, whereby the air moves from high-pressure areas to low-pressure areas (Figure 2.13).

The precise form of the wind profile is also influenced by the shape of temperature and humidity profiles. When wind streams

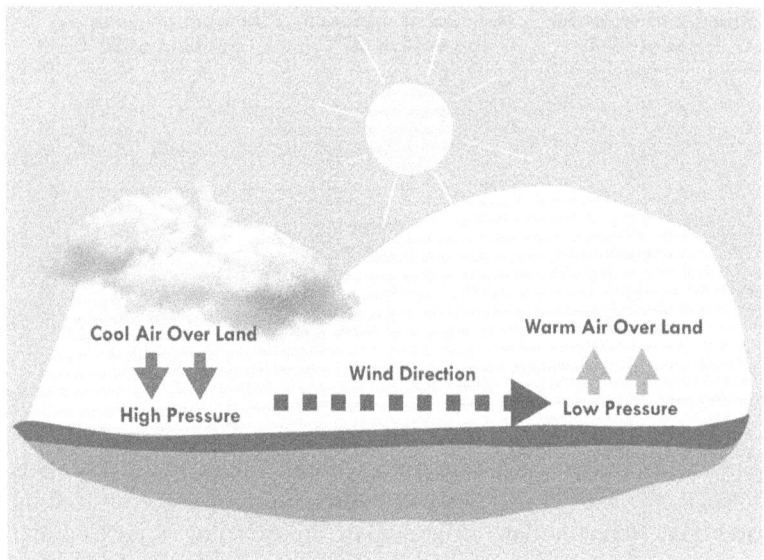

Figure 2.13 Wind formation resulting from high- and low-pressure systems

parallel the ground surface, the flow is laminar. However, on a smaller scale, turbulence is generated by the surface's irregular shape and aerodynamic roughness (Hutjes, 1996) (Figure 2.14). Changes in the landscape make the wind flows unstable and breaks them down into chaotic eddies. The more heterogeneous the landscape's surface, the more complicated its wind flow system can be. Therefore, local wind flow is highly irregular, with rapid velocity and speed variations (Barry and Blanken, 2016). On a farm-level scale, for example, wind can be funnelled through corridors or blocked through the presence of vegetation or shelterbelts.

Wind plays an essential role in the exchange of sensible heat and matter between the surface and the atmosphere. The wind 'wipes out' the boundary layer where these exchange processes occur, and water vapour is accumulated (Stoutjesdijk and Barkman, 1992). According to Michael Allaby (2015), this is called the 'clothesline effect' as wind dries clothes on a line by sweeping away the moist air around them and replacing it with drier air, increasing the evaporation rate.

Both evaporations of soil moisture and transpiration from leaves increase with wind velocity (Yoshino, 1975). The drier the atmosphere, the stronger this increasing effect of wind on evapotranspiration: wind can only replace saturated air with slightly less saturated air in more humid conditions. Thus, the wind speed affects the evapotranspiration rate to a far lesser extent than in arid conditions. In dry conditions,

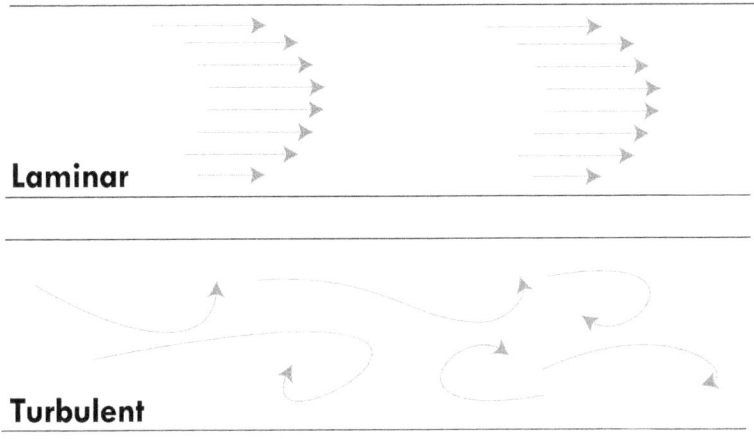

Figure 2.14 Turbulent and laminar flow. When the wind flow is parallel to the ground surface, it is called a laminar flow. When the wind is irregular because of the surface roughness, it is called a turbulent flow

slight variations in wind speed may result in more considerable variations in the evapotranspiration rate (Allen et al., 1998).

Local winds also play an essential role in dew formation as light winds help dew formation in unsheltered sites. According to Monteith (1957), dew occurs by turbulent transfer and is, in fact, negligible when winds are less than 0.5 m/s. Here, the wind is essential to replenish moist air. However, too strong winds (> 4.57 m/s) inhibited dew formation. Significantly when the humidity deficit increases, dewfall becomes more sensitive to increasing wind speed, similarly to the process of evapotranspiration, as explained earlier.

2.7 The local climate system

The previous sections deconstructed the local climate according to all six climatic components: solar radiation, soil moisture, soil temperature, air temperature, air humidity, and wind direction and speed. Together these components work as a system and form the local climate in which humans live and plants grow.

This behaviour of the local climate systems is of utmost importance for agriculture as together they shape the climatic conditions in which crops can thrive. Either directly or indirectly, all six components influence processes and factors essential for crop production.

An overview of the processes relevant to farming practices and plant growth is provided in Figure 2.15. It should be noted that this is a simplified representation of the climatic system.

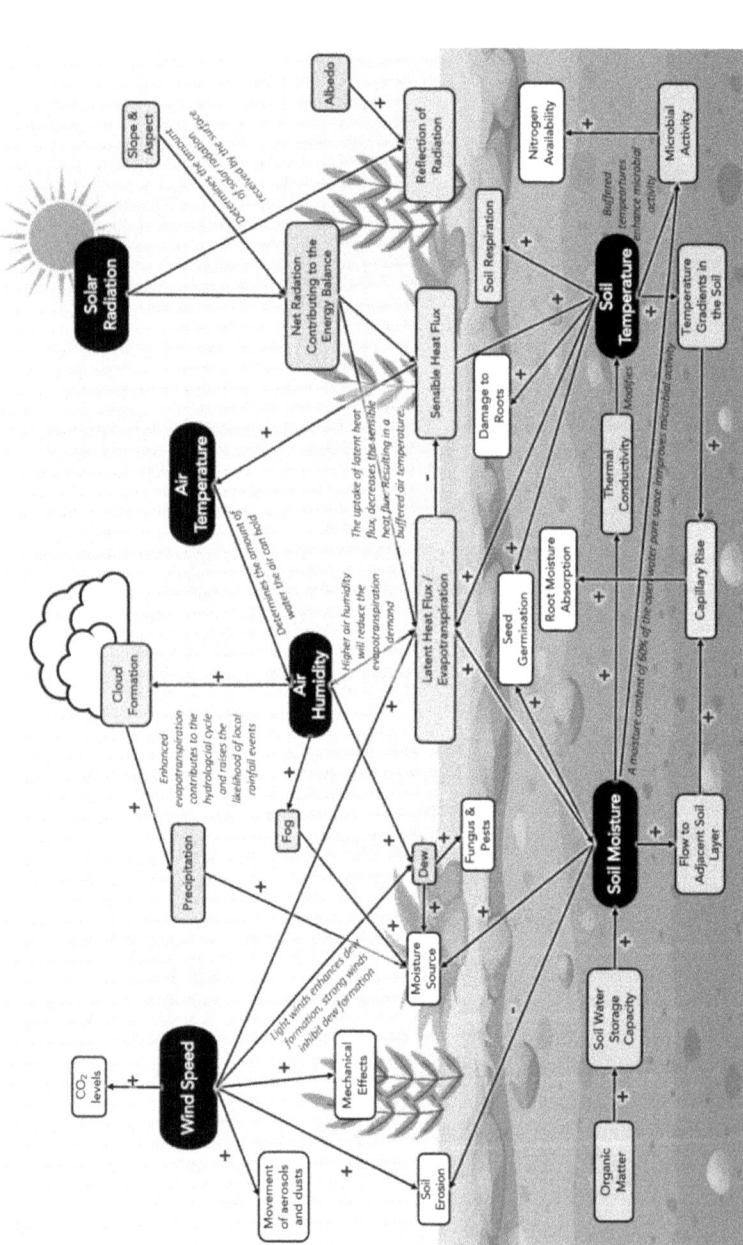

Figure 2.15 Conceptual model of all local climatic components and their interactions relevant for agriculture. In the black boxes, the local climatic components are displayed. In the white and grey boxes, the processes (grey) or factors (white) that influence crop production are shown. The arrows and the signs indicate their relationships: a '+' indicates a positive (increasing) effect, and a '–' indicates a negative (decreasing) effect
Source: MetaMeta

The positive news is that we can impact all these factors and processes, enhance crop production, and improve local climate conditions by making changes on a site. Moisture conservation, shading, regreening, windbreaks, and agronomic measures alter this local climate system. When done right, we can make a farm or landscape thrive and improve and buffer crop production conditions considering climate change.

CHAPTER 3
Improving the local climate: start of a guiding kit

> *There is a temperature difference. Due to the trees that grow in the revegetated area, it is much cooler. It also makes us happy to be around the greener area*
> Farmer in the Raya Valley, Tigray, Ethiopia

> *My farm is located on the Yatta plateau in Kenya. Yatta is known for her dryness and harsh farming conditions. However, it feels cooler after planting trees on my farm, and there are less dust storms. It almost feels as if I am no longer on Yatta anymore!*
> Josephine, Farmer in Kwa Vonza, Kenya

> *We were able to control the stem borer in paddy by controlling the local climate. We temporarily drained the land and changed the air humidity. Together with other measures this reduced the extensive damage caused by the stem borers. With water and climate management we can control many agricultural pests*
> Dr Ferdous, retired Principal Agricultural Extension Officer, Bangladesh

This chapter aims to build up a systematic approach to preserve and improve local climates by developing a guiding kit. This guiding kit gives practical ideas on deploying land and water management measures to manage the local climate from plot to landscape level. Because of the intricacy of the local climate (see Section 2.7), influencing one aspect sets in motion changes in many other aspects.

Most of the local climate measures discussed in this chapter have multiple uses. In addition to what they bring to the local climate, they are usually productive assets in their own right: a forest, a water body, a soil management practice, a terraced landscape, a series of hedgerows. What is required is to use these interventions in a more considered manner to craft better local climatic conditions. As such this differs from ecosystem-based adaptation in that the main climate parameters in a local landscape are directly influenced to create more conducive and more stable conditions, in terms of air humidity, soil moisture, local temperatures, winds, and even to a modest degree rainfall.

In improving the local climate it is essential to understand the current local climate and to appreciate what affects it and particularly what threatens it. Converting forests, opening up roads (Box 3.1), changing agricultural land management, and diverting water streams can all negatively affect local climates and undermine their resilience. Managing local climates is as much about preserving what is valuable as taking proactive improvement measures.

An improved local climate will support crop production. Figure 3.1 displays how farmers and watershed managers can use local climate management to improve agricultural yield.

Managing the local climate is both a science and an art. There is no standard plan or list of techniques that can be replicated across every site. Local conditions, topography, soils, hydrology, economy, ecology, and farming systems differ from place to place. Every location is unique, and every area has its own best approach. What is important is to be pragmatic and adaptive:

1. Carefully observe and understand the local climate as it is now, across the day, across different seasons, and in times of stress, and understand the different components that determine this local climate. Chapter 4 discusses several approaches and tools for measuring and monitoring the local climate.
2. Conceptualize desired improvements in microclimate and mesoclimate for the given area, in terms of the different parameters (temperature, wind, moisture, humidity) and the overall buffering effect (during the day or across the year).

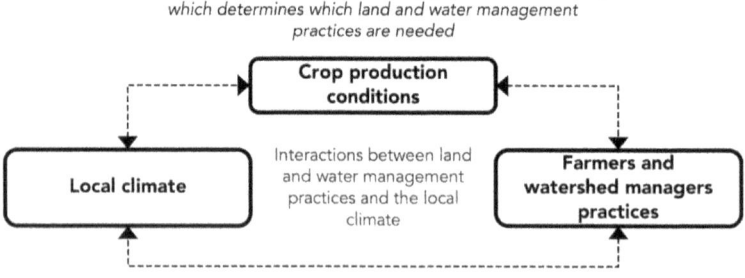

Figure 3.1 Visual representation of human–nature interactions between farmers and watershed managers and the local climate

Box 3.1 Harmful effects of road development on the Nepalese mountain local climate

The example of road development in the mountains of Nepal demonstrates the often-unintended and dramatic harmful impact on local climates of major landscape changes.

Road construction in the mid-hills of Nepal has led to openings in the steep mountain terrain. This increases the exposure of the earth to sunlight and wind, changes runoff patterns, and causes areas to dry out further, unavoidably changing the environment of mountain areas. Combined with reduced tree canopy, these changes significantly impact the local climate, resulting in less water retention, moisture loss, increased temperature, and general desiccation. The drier local climates and reduced soil moisture affect the soil temperature and microbiology. This means more heat exposure and less microbial action in the soil, which will reduce the capacity to fix nitrogen, with repercussions for vegetative growth. Because of this drying, the moderating effects will unravel, daytime temperatures will increase, while night-time temperatures will drop even more. These changes may also affect local rainfall patterns and the occurrence of dew, which is an essential source of moisture for vegetation.

This example shows that it is crucial always to consider the effect of landscape alterations on the local climate and minimize negative impacts. In this case, this can be done by implementing practices that retain moisture and regreen the area next to the roads. The large quantities of rocks and boulders that will become available during the construction of roads can be used to implement moisture-retaining measurements such as eyebrow terraces, rock bunds, or stone strips. These will slow down runoff, increase groundwater recharge, intercept sediment, and build new soil layers (Van Steenbergen et al., 2021).

Photo 3.1 Road openings interrupt subsurface water flows and increase exposure to wind
Credit: MetaMeta

Examples of desired improvements could be to slow down high wind speed on the farm, increase temperature buffering capacity of a landscape, or increase humidity and moisture levels in a watershed.
3. Understand the likely local climate components that, when preserved and/or altered, may bring benefits to several functions, including crop production. For example, if a farm has very strong wind blowing over the field, slowing down this wind speed might lower evaporation from the soil and lead to improved soil moisture levels.
4. Check the interventions in this chapter and see if they are applicable in the given context. When not applicable, think of alternative measures using the same logic. We can use the table presented at the end of this chapter to see what landscape characteristics are required for the practices presented in this chapter and the order of magnitude of the impact (Table 3.1). Beyond the interventions in this chapter, there may be others to be considered.
5. Consider the positive and negative effects of the introduced interventions on farming operations, the local climate, and the farm economy.
6. Test, implement, and evaluate the results. For this, the monitoring approaches and tools presented in Chapter 4 can be used.

Box 3.2 shows a practical example of how the above-mentioned steps are applied on a farm.

The local climate and the practices that alter it are influenced by and affect a myriad of processes such as nutrient cycling, water availability, biodiversity, and crop production (see Figure 2.15 in Chapter 2). Therefore, we should perceive every farm and landscape as an agroecosystem driven by interconnected ecological processes. Central to this perspective is to work with natural systems rather than against them, resulting in more resilient and productive local ecosystems where natural resources are enhanced. As mountain farmer Sepp Holzer states in his book *Desert or Paradise*, a farmer should always ask her or himself, 'Are my actions part of the natural cycle, or am I the disturbing element?' (Holzer, 2012: 32). Sustainable agriculture is considered a system where people seek to optimize productivity and at the same time maintain a healthy environment for present and future generations (Antle et al., 2017; Peterson et al., 2018). This perspective aligns with the 10 agroecological principles, introduced by the Food and Agriculture Organization of the United Nations (FAO, 2018): diversity; co-creation of knowledge and sharing; synergies; efficiency; recycling; resilience; human and social values;

> **Box 3.2** Local climate management at the farm
>
> The steps to introduce local climate management at a farm level are better clarified with an example:
> A family realizes that their farm is suffering from poor crop production because they cannot meet the ideal water requirements of the tomato fields (Step 1). They want to increase the soil moisture levels to enhance the transpiration of crops as they suspect that too much water is lost through evaporation, accentuated by rising summer temperatures and dry winds (Step 2). They identify wind speed and directions, too high air temperature, and low air humidity as possible causes of excessive evaporation (Step 3). By looking at the available options, they first decide to control the desiccating southerly winds. Tree planting would be an ideal option, but it would take time to become effective. Accordingly, they choose to also create a small fence of fabric around the tomato plot to shelter it from the wind. The fabric fence will be removed when the trees are fully grown (Step 4). Considering farming access, they decide to keep an opening in the windbreak to access the farm easily. They also decide to integrate fruit trees in the windbreak to reap extra benefits (Step 5). The family chose to leave a small portion of the farm without any form of wind protection and to compare it with the protected farm. If it works, they will expand the windbreak to the whole farm (Step 6).

culture and food traditions; responsible governance; and circular and solidarity economy.

As the horizontal scope of the local climate ranges from a few metres up to many kilometres (Table 1.1), the scale on which interventions can impact the local climate varies. Interventions can improve the immediate environment of single plants, but measures such as regreening and water retention can change the local climate at a landscape level. These two levels are closely intertwined. In other words, there is a 'nesting' effect of microclimate and mesoclimate measures, and at the same time, the landscape mesoclimate defines the boundaries for farm-level microclimate.

This chapter discusses several practices and how they influence the local climate. The practices are diverse. Some transform the microclimate at farm levels; others address entire landscapes. The measures have worked to mitigate very high or very low temperatures, modify wind patterns, affect soil moisture or air humidity, and even harness local rainfall, thus driving various local climate changes.

The practices are categorized into five groups, discussed below: the management of water and moisture (3.1), adjusted agronomic measures (3.2), the management and preservation of vegetation (3.3), solar radiation management (3.4), and wind management (3.5). The last section (3.6) of this chapter brings these different groups of practices together and summarizes the evidence on how they affect local climate parameters. Beyond the practices discussed in this chapter, others are waiting to be documented and systematically applied. There is a wealth of local practices whereby different elements of the local climate are influenced to sustain crop production and other ecosystem services in challenging circumstances.

3.1 Management of water and moisture

There is an extensive repertoire of water management and moisture conservation measures that all have a unique bearing on the local climate, depending on the nature of the measure, location, and spatial arrangement (number and density). The way water and moisture are retained in a landscape in space and time is a central organizing force in local climates.

The essential source of water in most locations is rain. Rainfall reaching the ground goes through a process called partitioning. A portion of the water infiltrates the soil and aquifers, and a certain amount ponds up in natural depressions or artificial storage. If the ponding capacity is exceeded, the water will flow overland along slope gradients. Initially, the flow is laminar and forms a shallow sheet of slowly moving water over the soil surface, but over time it gathers speed and concentrates in flow paths. These paths can be natural waterways, rills, and gullies carved anew in the land by the erosive force of the flowing water or artificial channels. Different types of water conservation practices can influence the runoff:

- Retaining the laminar flow behind simple barriers to enhance in situ infiltration. These are continuous structures perpendicular to the water flow: check dams, trenches, micro basins, swales, zaï pits, preferably laid in systematic consecutive lines and connected, concentrating moisture retention and groundwater recharge.
- Constructing barriers such as check dams and gully plugs in ravines and gullies, as well as irrigation or floodwater diversions from local rivers and streams, to concentrate the runoff.
- Storing runoff and rainwater in water reservoirs to make it available over time, while often serving other functions, varying from aquaculture to irrigation or hydropower.
- Entire landscape moisture transformations in the shape of bunding and developing terraces on sloping terrain. An example of this is road water harvesting, whereby road infrastructure is used to improve the landscape.

The proposed measures may vary in shape, dimensions, and location. Still, they all share a common effect on the local climate of the farm by increasing the water retention in the landscape and concentrating

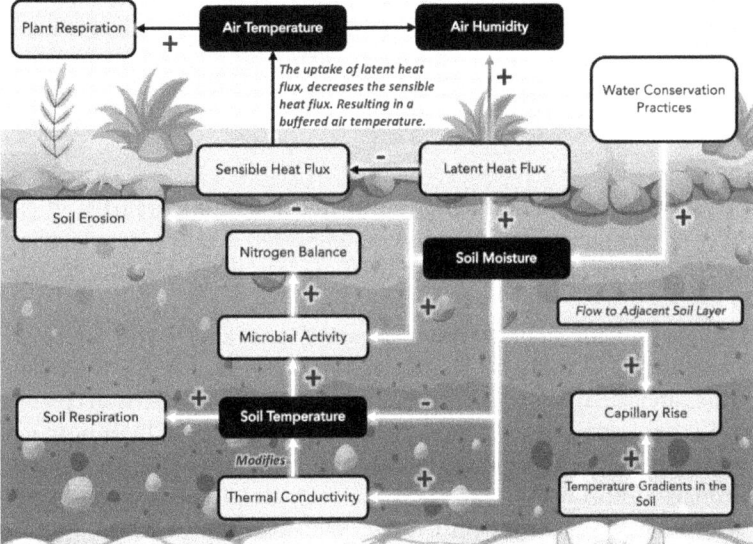

Figure 3.2 Overview of the daytime influence of water conservation practices on the soil temperature, air temperature, air humidity, and soil moisture interactions. A '+' indicates one component having a positive (increasing) effect on the other component and a '–' indicates one component having a negative (decreasing) effect on the other component
Source: MetaMeta

this in certain parts of the landscape. Some of the water is held in the upper layers of the soil. Some may percolate to a deeper layer, eventually replenishing shallow groundwater and enabling capillary rise outside the rainy season.

Increased soil moisture levels influence air humidity, soil temperature, and air temperature. Figure 3.2 shows an overview of the effect of increased soil moisture levels, as was discussed in Chapter 2. Surface water storage influences air temperature and humidity and can have mild effects on local winds too.

Without being exhaustive, this section describes a number of water and moisture management practices.

Harvesting and retaining sheet flows

A very effective way to retain water in a landscape is to 'harvest' the rain runoff and intercept the sheet flow. This ensures that entire areas become wetter and with it, the entire local climate is transformed, besides the other benefits from such water harvesting. There are many techniques, all differently suited to local conditions. Some focus on creating moist conditions for plant growth, and others also facilitate groundwater recharge. Below, several such methods to harvest sheet flow are presented.

Check dams
A check dam is a small barrier constructed across a drainage ditch, gully, or channel to lower runoff water speed. A check dam may be built with various materials such as stone, sandbags, gabions, or logs (Photo 3.2).

A check dam reduces water scouring and allows the retention of sediments. A study by Nyssen et al. (2010) on a north Ethiopian water catchment shows that the annual runoff coefficient decreased by 81 per cent due to check dams. This reduction in runoff led to a rapid recharge of the groundwater table after the dry season. Norman et al. (2015) researched the impacts of check dams in the Chiricahua Mountains, south-east Arizona. They also found a lower runoff response to precipitation and higher flow volumes (28 per cent higher) supported by increased groundwater levels. This shows how check dams can increase baseflow in a catchment. A study by Castelli et al. (2019b) on the effect of jessour, traditional water harvesting check

Photo 3.2 Check dams in Ethiopia
Credit: Francesco Sambalino

dams widely adopted in the arid regions of Tunisia, showed that jessour sites, compared to non-jessour sites, can reduce olive trees water stress both in the humid and dry seasons.

Bunds

Soil bunds are low (<1 metre) continuous embankments that are constructed on farmland along contour lines (Photo 3.3).

Multiple structures are built in series at a set interval to allow all water generated in between lines to be retained by the following bund. They help maintain runoff distribution and diminish its erosive power. Soil bunds in Ethiopia reduced the surface runoff by 17–94 per cent (Herweg and Ludi, 1999; Amare et al., 2014). The runoff water is kept behind the bunds and can seep into the soil, increasing soil moisture levels. If the space behind the bunds fills up, the water can overflow to the lower tier of bunds or is discharged to a safe drainage waterway. Nyssen et al. (2007) found that the soil moisture storage was increased at both sides of the bund. They also showed that at a depth of 1–1.5 metres, soil water content could be increased by 5–10 per cent for at least two months after the rainy season. Another study on soil bunds also showed a reduced runoff after implementing this intervention (Taye et al., 2013).

Photo 3.3 Soil bunds along contour
Credit: Mathias Gurtner

Photo 3.4 Stone lines combined with zaï pits in Burkina Faso
Credit: MetaMeta

Bunds can be constructed with excavated soil, stones, or a mix of the two. On gentle slopes, bunds can be lower and more widely spaced. When stones are locally available, bunds can be replaced by simple stone lines (Photo 3.4) that still play a similar function.

When bunds are built or reinforced with stones, they also impact the farm climate through the high heat capacity of the stones: when the incoming energy during the day is high, this warms up the stones and makes stone bunds a heat source at night, helping prevent nearby crops from night frosts. Using light-coloured stone walls results in extra short-wave radiation reflection on nearby crops (Oke, 1995). Bunds can also lower wind speed when perpendicular to the prevailing wind direction. The windbreak section at the end of this chapter will elaborate more on the effect of wind reduction on the local climate.

Microbasins

Microbasins are small temporary water buffering structures built in high numbers over a given tract of land. Unlike bunds and terraces, they are not continuous structures, and therefore they need to be placed wisely to maximize water retention. They are staggered on consequent contour lines to allow the lower structures to capture the water that is not captured by the basins in a higher position.

Examples of microbasins are eyebrow terraces, zaï pits, and semi-circular bunds. Eyebrow terraces are circular and stone-faced structures, applicable in steep and degraded hillsides and community closures

Photo 3.5 Eyebrow terraces
Credit: MetaMeta

(Farahani et al., 2016) (Photo 3.5). A planting pit is commonly excavated at the centre to augment the water buffering capacity and establish a tree. A reduction of surface runoff of 19 per cent due to eyebrow terraces was found in Pakistan (Hussain and Irfan, 2012). An impressive example of how these structures are benefiting farmers in Lanzarote, Spain, can be found in Box 3.3.

Zaï pits are planting pits typically 15 centimetres deep and 40 centimetres in diameter (Shaxson and Barber, 2003). They are constructed during the dry season by digging out the soil and placing it on the downslope side. When the rains begin, the runoff from the soil surface upslope runs into the pits.

Semi-circular bunds are half-circle shaped basins excavated in the ground surface and surrounded by a semi-circular earth bund with the wingtips on the contour level (Mati, 2006). The objective of these microbasins is to harvest water and build up soil moisture levels (Photo 3.7). Adekalu et al. (2009) found that semi-circular bunds retain an additional 20 per cent of the runoff over the land in research done in the savannah belt of Nigeria.

When built in a high number over the landscape, they help reduce runoff water volumes and speed. Consequently, they also reduce soil erosion, making these practices important in maintaining soil

Box 3.3 Lanzarote, Spain: wind protection by half-moon-shaped bunds

The farmers of the Lanzarote Island of the Canary archipelago turned the ancestral punishment of volcanic forces and drought into a unique socio-cultural scenery: 'La Geria'. La Geria is now a fertile wine-producing terroir scattered with half-moon-shaped stone bunds that protect the vines from the trade winds blowing incessantly from the north-east during the summer months (Photo 3.6). Making this grape cultivation happen was an act of extreme adaptive capacity to work with nature to maximize agricultural land productivity under the harsh environmental conditions of the island. With an average of 100 mm per year of rainfall, water and wind are limiting factors of agriculture in Lanzarote. However, the Geria cultivation technique enables the thriving of vines despite this water scarcity and extreme wind. The vine plants are planted in inverse cone-shaped pits protected by half-moon stone bunds orientated against the dominant winds. This way, wind protection is maximized, and unproductive evaporation is minimized. Maintaining these highly humanized lands is delicate, labour-intensive work as farmers must do everything by hand or, traditionally, with the assistance of their primary farming tool: the camel. Periodically, farmers have to re-shape the pits and raise the volcanic sand as it moves downwards along the pit walls and submerges the base of the plant. For a good reason, farmers of Lanzarote are referred to as gardeners and caretakers of the landscape (Borgia, 2017).

Photo 3.6 Half-moon stone bunds that protect vines from strong winds in La Geria, Lanzarote

moisture and quality (Gedamu, 2020). The above examples show that a high density of runoff retention measures can significantly increase soil moisture, thus laying the basis for a better-buffered local climate.

Photo 3.7 Example of water harvesting through semi-circular bunds in Pembamoto, Tanzania
Credit: Justdiggit

Increased moisture levels and soil quality will also increase vegetation (Photo 3.8). Both interventions and similar variations thus affect the local climate as described at the beginning of this section (Figure 3.2).

Trenches and swales

Trenches are excavations (approximately 1 metre deep) constructed along the contours to collect and store rainfall water (Photos 3.9 and 3.10). Due to their higher storage capacity than bunds, trenches can withstand more increased water flow and are less prone to breakage. For this reason, they are often preferred when the slopes are steeper. The trenches are commonly segmented to ensure that water is homogeneously distributed across the structures dug on a contour. This rainwater harvesting system enhances soil moisture levels and increases water availability for a prolonged time after the rainy season. Kaushal et al. (2021) found 16 per cent higher soil moisture levels than on a site without trenches.

Swales are wide channels with a gentle cross-section used to either retain water in the landscape or gently discharge excess water while allowing a portion to seep into the ground. They are often kept with a

Photo 3.8 Example of increased vegetation as a result of increased soil moisture levels and soil quality in the semi-circular bund
Credit: Justdiggit

Photo 3.9 Trenches
Credit: MetaMeta

Photo 3.10 Trenches placed on the contour of a slope adjacent to Lake Ziway, Ethiopia
Credit: Simon Chevalking

grass cover, and if water is allowed to flow to a discharge outlet, small check dams may be built inside it (Ekka et al., 2021). By increasing the soil moisture levels, swales and trenches moderate local climatic temperatures, as described previously.

Terracing
One step from retaining rainfall and intercepting run-off is the reconfiguration of sloping landscapes through the development of terraces. Terracing changes the local topography dramatically and transforms the local climate. Soil movement is necessary to form consecutive benches of flat land separated by walls (risers) that make a terraced field profile resembling a staircase (Photo 3.11). Soil is excavated in strips and heaped to form the flat benches where cultivation occurs. The soil profile is much deeper on the benches than on unterraced land, which guarantees space for root development and a higher soil moisture storage capacity. Making terraces may seem labour demanding, but much can be achieved over one season when done collectively. In some areas, mini bulldozers are deployed for the purpose.

The benches are typically designed with vertical intervals ranging from 1.2 metres to 1.8 metres (Mati, 2006). By reducing the plot's

Photo 3.11 Agricultural terraces and grazing fields near Foz d'Egua, Piodao, Portugal

steepness, terracing helps buffer runoff water and decrease soil erosion while supporting soil water recharge (Mesfin et al., 2019). Runoff reduction and soil water recharge increase the soil moisture levels of a site – the flatter the terrace, the higher the water harvesting capacity. A reverse-slope bench is even more effective in capturing runoff (Mati, 2006). Mesfin et al.'s (2019) study on soil water conservation variations in Ethiopia's Tigray region showed an increase of 110 per cent in soil water content between non-terraced control plots and terraced ones. A study in the Fateh Jang areas of Pakistan found a 16 per cent increase in soil moisture content due to terrace structures (Rashid et al., 2016).

By altering the slope, terracing affects solar radiation absorption. The net increase or decrease in solar radiation on the surface differs significantly depending on altitude, latitude, slope aspect, angle, and season. For instance, terracing will confer a net direct solar radiation gain of 15 per cent on south-facing 30-degree slopes at the equator; however, it will reduce net annual direct solar inflow by 21 per cent on south-facing 30-degree slopes at 45-degree N latitude (Evans and Winterhalder, 2000). For low latitudes, terracing can thus optimize the use of solar radiation and, as a result, also increase the soil temperature as more solar radiation is received (Onwuka and Mang, 2018). However, the increased temperature may be buffered by increased soil moisture levels.

> **Box 3.4** Lomas de Mejía, Peru: fog collection for ecosystem restoration
>
> In the drylands of Peru in the late 1990s, a project was launched to rehabilitate the area of Lomas de Mejía in Peru. The aim was to reintroduce vegetation on the Lomas ecosystem, and this was done by trapping fog using artificial fog collectors. These collectors consist of a metal frame and are covered with a synthetic mesh, replicating the effect of leaf surfaces. The water collected was used to feed trees. Small vegetation and local ecosystems were sustained after two years of artificial fog collection (Semenzato et al., 1998). The habitat conditions, both in terms of diversity/frequency of plant and animal populations and plant cover, once kick-started, developed substantially over the years (Correggiari et al., 2017).
>
> Fog collectors can be found along several coastlines in Chile, Peru, the United States, Morocco, South Africa, the Dominican Republic, and Spain (Canary Islands) as well as in inland locations, such as Ethiopia, Guatemala, Yemen, and Tanzania. By harvesting water and increasing the soil moisture levels in an area, they created small local ecosystems with their own local climate.

Also, within a terraced landscape soil temperatures may vary. Based on thermal imaging, Tucci et al. (2019) established that the temperature variations of internal rows are 2°C lower in the morning. This was explained by the solar position and the effect of the terrace risers that provide a shading. From the early afternoon, this difference was evened out.

All the above-mentioned practices are designed to harvest surface sheet flows. However, another source of water is water directly from the air in the form of fog. Box 3.4 provides an example of how fog collection in Lomas de Mejía, Peru, improved an entire ecosystem.

Water diversions

Water diversions are systems of structures and measures that intercept or divert water into or from a waterway. This can be done using instream barriers such as dams, weirs, culverts, canals, or pipes that divert an excessive amount of water to prevent flooding or use this water for other purposes, such as agriculture. Altering the waterways' natural characteristics and spreading or diverting the water to another area impacts the local climate. This section discusses irrigation and floodplain farming.

Irrigating an area has dramatic effects on the local climate. As irrigation increases soil moisture, it can modify local energy balances and have a cooling effect. Research in a semi-arid irrigated region (the Columbia Basin Project) in the Pacific Northwest of the United States showed cooler land surface temperatures by as much as 20°C over the irrigated areas compared to nearby non-irrigated areas.

This effect was due to the direct (increased soil moisture and evapotranspiration) and indirect (land-cover and vegetation change) effects of irrigation. The greatest daily cooling thus occurs directly over the irrigated area. However, winds also play a role in spreading the irrigation cooling to surrounding areas. For light and variable winds, the cooling extended to areas 5 to 10 kilometres away from the irrigated region. Stronger winds allowed the irrigation-cooled air to travel up to 40 kilometres (Lawston et al., 2020). Increased soil moisture levels thus not only have a local climate regulating effect but might also lead to temperature buffering effects downwind. As shown in Chapter 2, soil moisture is also the first 'link' in the process chain that relates changes in soil moisture to clouds and precipitation. As described earlier, when a large area is irrigated, the surface air temperature is cooled. This increases the surface pressure, which is speculated to reduce rainfall over the irrigated land while enhancing rainfall in areas downwind. Alter et al. (2015) found that this phenomenon also occurs in the East African Sahel around the Gezira Irrigation Scheme. This local climate regulating effect of crop and landscape evapotranspiration should be considered simultaneously with the water consumption footprint of crop production.

Another example of the cooling of large water diversions comes from flood-based systems that use short-term flood situations to cultivate crops. In floodplain farming systems, either receding or rising floodwaters provide water for crop cultivation. Diverting flood water and or streams to a landscape and using and retaining it in a landscape has a significant impact on the local climate of that area, especially the temperatures. Tonolla et al. (2010) quantified this thermal heterogeneity for Alpine river floodplains. The maximum daily surface temperature for exposed sediments was 23°C, whereas, for the aquatic habitat, this was 11°C, highlighting the temperature buffering capacity of the water compared to the exposed sediments. This shows water diversions' impact on the surface temperature by steering the water through a landscape. However, it should be noted that diverting and retaining water upstream leads to less water availability downstream and a shrinking of water bodies from which the water is diverted (Soomro et al., 2020), which may undermine downstream local climates.

Water storage

Water retention ponds have a distinct effect on the local climate. In all geographies, water ponds affect the local climate, with the effects varying with the season and the specific circumstances. The main effects are on air humidity, temperature, and wind patterns.

Water ponds increase air humidity, as witnessed in the shape of fog and dew. This may sometimes sustain crop production, as around lakes where wetter conditions make it possible to grow crops. It can also have adverse effects, such as when dew 'burns' young vegetable seedlings in combination with the morning sun. Both effects have been witnessed from newly created storage ponds. The humidity effects are very much a function of the surface area of the local water bodies. Per volume of water, a shallow pond or lake has a larger impact on air humidity as there is more surface from which water can evaporate, also because they heat up more rapidly. The evaporation from shallow water bodies is higher than from deep water bodies.

The other effect is on local temperatures. Water ponds have a moderating effect on the temperature in the surrounding area – reducing high and low extremes. Box 3.5 provides an example of how water bodies are used against night frost damage in the Andes.

The effects of water ponds differ according to the type of weather, the seasons and the year, with the sphere of influence either widening or decreasing. These effects were empirically measured in the Brenne in France, an area dotted with many small water bodies (Nedjai et al., 2018).

Box 3.5 The use of water bodies against night frost damage in the Andes

An ingenious form of ancient local climate management is found in the Andes, where some farmers have developed an inherited complex farming system adapted to local conditions. These systems have helped them sustainably manage harsh environments and meet their subsistence needs without depending on mechanization, chemical fertilizers, pesticides, or other technologies (Altieri, 1996).

Photo 3.12 Example of a raised field farming system from above

(Continued)

Box 3.5 Continued

From the Andes to Amazonia, raised fields known as *Suka Kollus* (Bolivia) and *Waru-Waru* (Peru) can be found (Photo 3.12). This is an agricultural technique in flood-prone mountain plains used by pre-Columbian societies (Lombardo et al., 2011). When the canals around the fields are filled with water from the rains, they provide moisture for crops. During droughts, moisture from the canals slowly ascends the roots through capillary action, and during floods, the furrows drain excess runoff (Altieri, 1996). These water bodies also serve as a temperature buffer against prevailing night frost. As stated in the previous chapter, water surfaces are poor reflectors and thus act as a good sink for solar energy (Rosenberg et al., 1983). The buffering effect is achieved by letting the sunlight heat up the water stored in the canals between the raised fields. The stored heat is released at night, thereby helping protect crops against frost (Figure 3.3).

Water in the canals absorbs the sun's heat by day and radiates it back by night, helping protect crops against frost. The more fields cultivated this way, the bigger the effect on the microenvironment.

The platforms are generally 13 to 33 feet wide, 33 to 330 feet long, and about 3 feet high, built with soil dug from canals of similar size and depth.

Sediment in the canals, nitrogen-rich algae, and plant and animal remains provide fertilizer for crops. In an experimemt, potato yields outstripped those from chemically fertilized fields.

Figure 3.3 Functioning of the raised fields
Source: Adapted from Boerma, 2013

On the raised beds, night-time temperatures may be several degrees Celsius higher than the surrounding region (Altieri, 1996). This heat release also generates moisture available as condensed water to the crops (Lhomme and Vacher, 2003). According to Roldán et al. (2004), the moisture raises the relative humidity by 3.3 per cent, compared to the surrounding plains. The soil temperature in the raised fields also shows lower temperature variability than a flat plain. Angelo et al. (2008) found the diurnal temperature range for the flat plains to be between 10.7 and 20°C, while in the raised fields, the temperature lay between 11.5 and 18°C (Angelo et al., 2008). These more moderate temperatures of raised fields' soils increase resilience against climatic extremes (Ismangil et al., 2016).

The temperature differences between the pond areas and the surrounding lands were found to be relatively consistent in the order of 4–5°C higher in the winter, and a decrease was found in summer, tapering off over the sphere of influence that extended from 100 to 200 metres around the water bodies. The moderating effects of the water ponds were caused by the energy consumption in the evaporation of the stored water or by heat exchanges with the atmosphere. The effects depend on the size and water volume of the water bodies and the shape of the surrounding terrain – flat or undulating.

In managing the local climate, water bodies may be planned systematically with important decisions on their size, depth, and distribution, as far as the local terrain allows. As the water bodies have a climate regulating impact extending up to 200 metres around them, it can be assumed that a larger amount of water bodies spread over a landscape has the most optimal effect on the local climate. The depth of the water bodies should be enough to prevent water shortfalls from maintaining optimal buffering results. The larger the water volume, the larger the protective buffer zone (e.g. lakes were found to protect from frost up to a distance of 5 kilometres (Louka et al., 2020)).

The buffering effect might also be enhanced when water bodies are designed parallel to the prevailing wind direction. During the day or summer, the wind cools down while blowing over the water and brings a cooling breeze to the downwind area. Following the same temperature buffering principle, water bodies can be a key factor in controlling frosts.

Water bodies, especially large ones, also impact local winds. This effect changes over the day. In the night and the morning, when the land temperature may have reduced below the temperature of the water bodies, a mild wind will flow from the water to the land. During the daytime, this is reversed.

The management of water and moisture thus can make significant improvements to a local climate in terms of buffering capacity. All management techniques described in this section – harvesting and retaining sheet flows, water diversions, and water storage – can be combined to create the most optimal effect for the local climate (Figure 3.4 and Photo 3.13). Box 3.6 provides an example of landscape-wide water harvesting in the Tigray region, Ethiopia.

3.2 Adjusted agronomic practice

Because agriculture defines land use to a major extent, the type of farming and the agronomic practices have a huge bearing on the local climate. Several agronomic measures may be deployed to influence the

48 MANAGING THE LOCAL CLIMATE

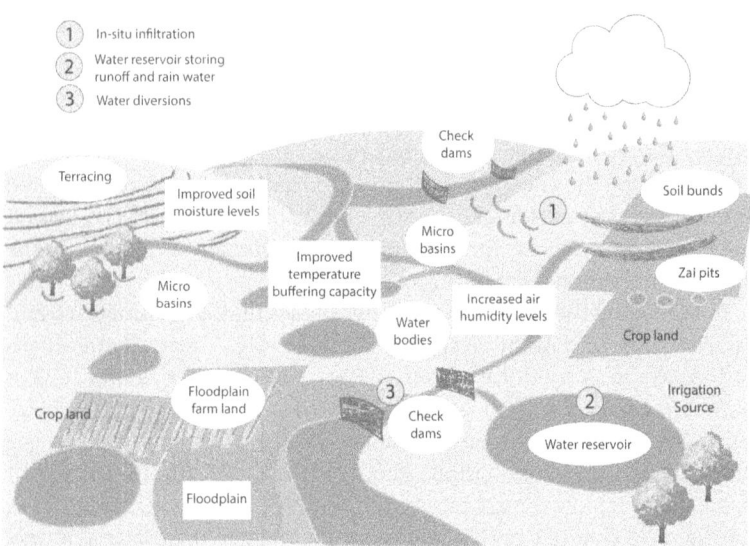

Figure 3.4 An overview of several water and moisture management techniques for local climate improvement implemented in a landscape
Source: MetaMeta

Photo 3.13 Water storage in abandoned borrow pit in Ethiopia

local climate, in addition to the primary function of crop production. The immediate impact of agronomic measures is very much in the domain of soil moisture and soil temperature – but as discussed in Chapter 2, soil moisture and soil temperature influence air humidity and air temperature and, subsequently, local wind patterns. The effect of the different measures may vary with the season, changing as bare land gets covered with crops. The agronomic practices discussed are soil management, tillage, mulching, intercropping and furrowing.

> **Box 3.6** Landscape-wide water harvesting in Tigray, Ethiopia
>
> Starting in the early 1990s, the government of Ethiopia implemented large-scale landscape restoration practices in the arid and semi-arid agroecosystem of the Tigray region (Photos 3.14 and 3.15). Entire watersheds were transformed by community action and local planning, using labour in the off-season. The repertoire of measures consisted of contour trenches, stone bunds, infiltration ponds, half-moons, and road water harvesting. The aim was to revert land degradation, increase soil moisture, reduce runoff losses, and increase agricultural yields.
>
> Castelli et al. (2018, 2019a) researched the effect of these landscape restoration and water harvesting practices on the local climate. By analysing remote sensing data, they found an increased capacity of the catchment to maintain soil moisture accumulated in the rainy season after the practices were implemented. Consequently, this reduced land surface temperatures up to 1.74°C in the four months after the rainy season. The implementation of landscape restoration and water harvesting measures thus resulted in a climate regulation effect in the watershed.
>
>
>
> **Photo 3.14** Example of landscape restoration and water harvesting measures resulting in increased soil moisture accumulation in the rainy season which results in a local climate regulation effect
> *Credit:* Giulio Castelli
>
> *(Continued)*

> **Box 3.6** Continued
>
>
>
> **Photo 3.15** Example of a pond in the restored landscape. During the day the sun warms the water up. During the night that heat is slowly released and, through dew formation and evaporation, the area around it is cooled and kept moist
> *Credit*: Giulio Castelli

Soil management

The soil composition is a major factor in determining the soil temperature and moisture content. The chemical composition, the particle size, and the organic material content play a role in this. Different soil types have various heat transmission capacities. Applying organic fertilizer or green manure changes these properties and modifies soil temperature fluctuations on daily and seasonal scales (Jeon et al., 2008; Zhang et al., 2013). Feng et al. (2021) conducted a study in the Hetao irrigation district in China and found a 2°C soil temperature reduction and reduced day–night differences after 30 tonnes/hectare organic fertilizer application. Kuzucu (2019) saw a 55 per cent increase in soil moisture after applying organic fertilizer in combination with soil bunds.

It should also be noted that the colour of soil amendments might alter the surface albedo. An example is the dark colour of the organic fertilizer biochar (charred biomass). The application of biochar results in substantial reductions in soil surface albedo, which leads to higher absorption of solar radiation (see Chapter 2). This increased albedo should be considered when applying dark coloured organic fertilizer (Meyer et al., 2012). Pale materials, such as kaolin or ash can increase soil surface albedo and decrease soil temperatures by reflecting more solar radiation. Applying these materials can thus be helpful in areas with high levels of solar irradiation.

Tillage

Different soil tillage methods have different effects on soil temperature. The main distinction is between land preparation with a mouldboard and disc ploughing on the one hand and conservation tillage (CT) on the other hand. CT is a general term to describe tillage where land preparation is minimized, the land is not disturbed, and crop residue is left on the surface. Both minimum tillage and no-tillage are forms of CT. A universal definition of CT does not exist, but experts commonly refer to CT as any form of tillage that leaves at least 30 per cent of crop residue to protect the ground (Carter, 2005).

CT practices decrease soil disturbance and improve soil aggregate stability. This improves water infiltration and thus increases soil moisture and reduces runoff and soil erosion, which benefits soil sediments and nutrients (Chen et al., 2011; Chai et al., 2014; Busari et al., 2015; Adimassu et al., 2019). A study by Blevins et al. (1971) in Kentucky found that soil moisture in the top 0 to 8-centimetre soil layer under no-tillage was 10–15 per cent higher than under conventional tillage.

CT also reduces soil temperatures compared to conventional tillage. This is attributed to thermal admittance differences, heat flux to a deeper depth, and total heat inputs to the soil profile, leading to a more moderate upper profile soil temperature (Johnson and Lowery, 1985). A study on rainfed Mediterranean vertisol showed differences between a no-tillage and a conventional tillage practice of between 0.7 and 2.6°C depending on the month of the year (Muñoz-Romero et al., 2015). Shen et al. (2018) also found that no-tillage practice led to a lower soil temperature up to 1.5°C. Higher residue coverage also caused lower soil temperatures. CT thus positively affects soil moisture and balances soil temperatures; it can therefore be used to improve the farm's climate.

Mulching

Mulching is the practice of covering the ground surface with an insulating layer to separate soil from the atmosphere. Mulching material may be of organic (crop residues, leaves, compost) (Photo 3.16) or inorganic (plastic sheets, gravel) (Photo 3.17) origin (Acharya et al., 2005).

Mulch forms a barrier to the heat and vapour flow between the soil and the atmosphere and consequently inhibits heat and moisture exchanges (Wilken, 1972; Rosenberg et al., 1983). Mulching decreases the evaporation rate, enhances infiltration (when permeable), and thus increases moisture conservation (Stigter, 1984a) (Figure 3.5).

Photo 3.16 Use of straw mulch for crop production

For example, Dhaliwal et al. (2019) showed that soil moisture was 4.2 per cent higher in sites covered with rice straw mulch than in un-mulched areas. This field experiment was conducted in Ludhiana, India, characterized by a semi-arid, subtropical climate with sandy loam soil. Adeboye et al. (2017) found that mulch increased the soil water storage in the upper 30 centimetres of the soil by 17 per cent. These increased soil moisture conservation levels are mainly the result of lower evaporation. Another way mulching can improve soil moisture levels is by enhancing the additional water supply of non-rainfall atmospheric (dew and fog) water into the soil. This requires mulch layers that do not take up and retain moisture

Photo 3.17 Use of plastic mulch for cucumber and strawberry crops in Urfa, Turkey
Credit: Sukru Esin

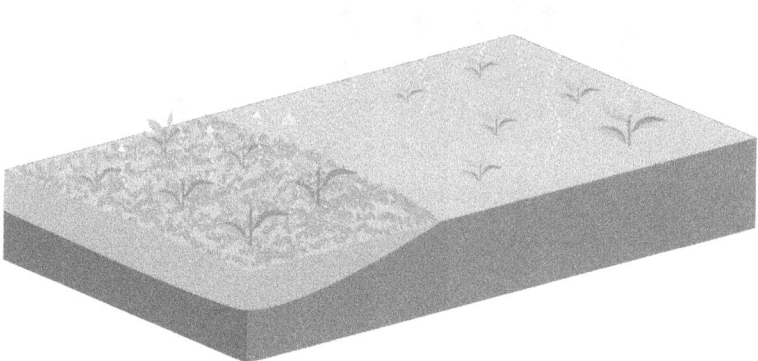

Figure 3.5 A thin layer of leaves (left side of the plot) acts as mulch that reduces the loss of soil moisture through evaporation

themselves; instead, they provide additional water by nocturnal condensation (Graf et al., 2008). Box 3.7 provides an example of dew harvesting in Harraz, Yemen.

The previous sections explain that increased soil moisture leads to soil temperature moderation. Mulching thus influences the soil

> **Box 3.7** Harvesting dew in Harraz, Yemen
>
> An example of spectacular modification of the local climate comes from Yemen. In Harraz, an entire valley – impressively – was blanketed with stones, while coffee trees grow in between (Photo 3.18). This stone blanket dramatic transformation minimizes soil evaporation and optimizes the capture of dew. The dew forms when the moist air touches the cold stones in the morning and drops below the dew point, that is, the point where air temperature can no longer sustain the moisture level (Van Steenbergen, 2013a). While some say that dew does not add up to much water, the law of many small numbers applies in Harraz, especially as the air close to the Red Sea is relatively moist.
>
>
>
> **Photo 3.18** Valley blanketed with stones and trees in Harraz, Yemen
> *Credit:* MetaMeta

thermal regime with a moderating effect, and it keeps the soil cooler during hot spells and warmer during cooler weather. This effect is also seen on diurnal temperature variations (Acharya et al., 2005; Kingra and Kaur, 2017). Mulching also reduces the incoming radiation on the soil surface, and by acting as a thermal insulator, this moderates the soil temperature (Gardner et al., 1999).

Kader et al. (2017) highlighted this buffering effect when they looked at the impact of different types of mulch on soybean production in Japan and showed that mulching buffered extreme soil moisture and temperature fluctuations. They found that the mulching treatments, compared to the control, lowered soil temperature by 2°C at a 5-centimetre depth and 0.5°C at 15- and 25-centimetre depths. Simultaneously they showed that plastic and straw mulching stored the highest quantity of soil moisture whereas bare soil stored the lowest. A study in Australia showed that surface temperatures fluctuated between 20°C and 50°C between night and day without mulched soil, and on the mulched sites the temperature only fluctuated between 20°C and 38°C (Bristow, 1988). Tayade et al. (2016) found that mulching increased soil moisture conservation from 0.70 to 5.92 per cent and buffered the soil temperature at 25.1–27.2°C in the top 5-centimetre layer of soil compared to a wider fluctuation in the control plot (26.9–34.0°C). This study was conducted in Coimbatore, India. More minor temperature fluctuations in the soil also favour root development. An example of how this temperature regulating effect of mulching is used to enhance wine production in Vinsobres, France is provided in Box 3.8.

Moreno and Moreno (2008) researched the effect of plastic material for mulching. The soil temperature reached under plastic materials is higher, which could be a disadvantage in hot climates and advantageous in cool climates. A common practice in northern China, for instance, is to cover furrows with a black plastic sheet to create a humid and warm environment for seeds to germinate. In other areas, white plastic mulch has been used to reduce the warming effect. In general, however, plastic mulch is not recommended due to its non-degradable residues. There is no effective practical biodegradable plastic mulch available that does not degrade too early (Van Steenbergen, 2013b).

Intercropping

Intercropping is the cultivation of two or more crops on the same field (Photo 3.20) for at least a part of the growing season (Stomph et al., 2019). Farmers can apply intercropping according to many designs, where crops, spacing, and timing are the main variables. Layouts differ. One of the crops may be planted at full density and intercropped with other crops at lower densities, or the first crop itself may be planted at a lower density to make space for the other crops (Keating and Carberry, 1993).

Intercropping is a time-tested practice that has always helped increase farm resilience. If one crop fails, they would still have a chance to harvest the other crops. In mechanized agriculture, this system became a niche. However, in some countries, it remained central. In Malawi and northern Nigeria, 94 per cent and 83 per cent, respectively, of all cropped land is under some form of intercropping (Knörzer et al., 2009).

56 MANAGING THE LOCAL CLIMATE

Box 3.8 Stone cover on vineyard in Vinsobres, France

Vinsobres is a medieval village in the Drôme in France. Farms here produce pine trees, cherries, and olives. However, the distinguishing element in the landscape of Vinsobres is the rolling and often terraced vineyards. In many vineyards, the soil is covered with stones acting as a mulch layer. In some cases, large rocks are crushed to create a thicker stone blanket (Photo 3.19).

Photo 3.19 Vineyard with a stone terroir in Vinsobres, France: the stones regulate the temperature in the vineyard, insulate the soil and retain heat in the soil, prevent evaporation of moisture from the soil, and enhance dew formation

The stones achieve several feats that are important for the vineyard's production. First, the stones regulate the temperature in the vineyards, insulating the soil and improving the local climate of the vineyards by retaining heat within the soil at night and preventing frost. Second, the stones that cover some fields help maintain the soil moisture by preventing moisture evaporation from the soils. On some colder nights, they create dew, ensuring a certain moisture level. This example shows how using stone mulch can create a local climate in which the vineyard thrives. In addition, the stone mulch adds to the distinct flavour of the wines, the so-called 'terroir'.

Photo 3.20 Intercropping of ginger and maize

Intercropping provides several advantages to growers, even though it might require a change and intensification of management practices. Many studies show that plant diseases, plant pests, and weeds are negatively influenced by intercropping, with a decrease of 79 per cent, 68 per cent, and 86 per cent, respectively (Stomph et al., 2019). The different types of crops also support and protect each other (Holzer, 2012). Furthermore, a well-designed intercropping can positively influence crop yields. The critical point is to maximize plant niches and decrease competition between plants.

Intercropping influences the farm climate, and it has particular effects on soil temperature and soil water content. Experiments in Kenya on a potato, lablab, and lima beans intercrop measured a significant decrease in mean soil temperature ranges (at 0–30 cm) between sole potato fields (22.6–28°C) and potato/legumes intercrop (18.1–23.7°C). Soil moisture was also positively affected with the legume/potato intercrop. During the short rains, the soil water content

> **Box 3.9** Intercropping to reduce the heat in Preah Vihear, Cambodia
>
> To grow vegetables in tropical conditions can be a challenge. Farmers in the vegetable zone in northern Cambodia deploy several practices to deal with high temperatures and even out temperature spikes. First is intercropping; combining two crops in one parcel has several benefits – pest control, increased soil fertility, additional produce – but it is also done to control land temperature. Common combinations are pumpkins with waxy corn or spring onion with leafy vegetables. A related practice is cover cropping. A promising method is to combine the cultivation of peppers with cover crops, such as *Arachis repens*. Though the practice is not so widespread, there are several benefits to using cover crops, besides regulating soil temperature: slowing down erosion, improving soil organic content, enhancing water availability, smothering weeds, helping control pests and diseases, increasing biodiversity, and attracting pollinators.

ranged between 2 and 18 per cent in the sole potato plot, whereas in the lima intercrop, the values recorded were between 9 and 26 per cent and between 7 and 27 per cent in the lablab intercrop (Nyawade et al., 2019). A decrease in water losses explains these effects through soil evaporation caused by increased canopy size and increased soil shading. Another promising example can be found in Sri Lankan rubber plantations. The expansion of rubber production intercropped with seasonal crops led to multiple benefits compared to the former monocrop production. Besides enhanced livelihoods, a significant improvement was found in the farm's climate. The tree canopy given by the rubber creates a high level of boundary layer resistance restricting water evaporation. Rubber cultivation reduced the air temperature by 3.7°C and relative humidity levels were 11.4 per cent higher compared to seasonal crop lands. This has provided better working conditions for farmers and would have an effect on the mesoclimate once the targeted extent of rubber is grown (Rodrigo et al., 2014). Box 3.9 provides another example of how intercropping together with cover cropping was used to reduce the heat in Preah Vihear, Cambodia.

Furrowing

Furrowing prepares the land with alternating furrows and ridges (Figure 3.6). Farmers can align the system to the contour lines to maximize water retention, or it can be gently inclined to favour water drainage when water is in excess.

When it rains, water accumulates in the furrows, resulting in improved drainage (El-Halim and El-Razek, 2014) and increased in situ water utilization efficiency (Gebreegziabher et al., 2009). Provided that the furrows have a gentle or nil side gradient, this practice reduces soil erosion induced by water. It spreads the water more evenly over

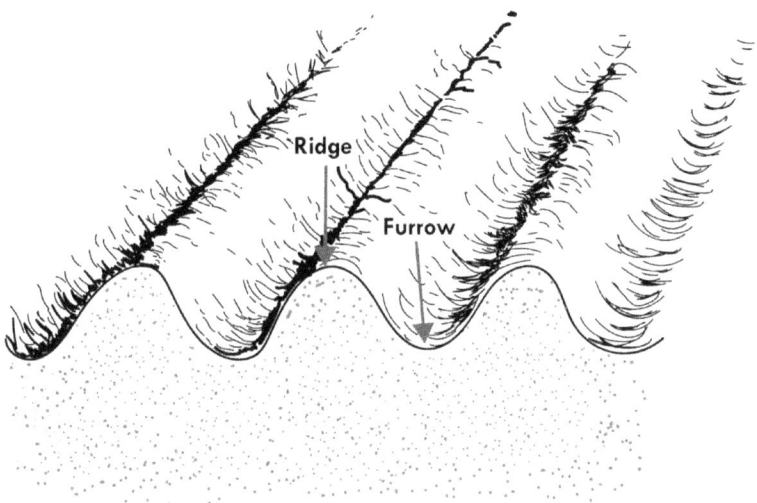

Figure 3.6 Top view and cross-section of furrows and ridges

a site (Farahani et al., 2016) with a consequent increase in water infiltration and soil moisture. Milkias et al. (2018) found an increase of soil moisture storage by 121.87 per cent on a plot in Oromia, Ethiopia.

Furrowing methods affect soil temperature as well. As the slope of a site affects absorbed radiation intensity (as was discussed in the section on terracing), ridging impacts the amount of solar radiation reaching the ground. The surface of the ridge that faces the sun receives more energy. By manipulating the ridges' geometry and orientation, a farmer may take advantage of the cosine law of illumination (as was discussed in Chapter 2), and better use of available short-wave radiation can be accomplished (Gardner et al., 1999). The sunlit surface may thus receive solar radiation at or near its local zenith. Ridge and furrow geometries constitute a radiative 'trap' for solar radiation and outgoing long-wave radiation (Oke, 1995). The trapping of solar radiation by day increases the maximum soil temperature and reduces surface cooling (Oke, 1995). This is especially useful in spring when heating is critical for germination and when the sun's elevation is still low (Figure 3.7).

However, studies by Burrows (1963) and Buchele et al. (1955) on soil temperatures in furrows, ridges, and a flat microtopography show that temperatures were generally highest in the ridge planting and lowest in the furrows because of the occurred shading by crops. Thus, this increase in received solar radiation by the surface and increasing soil temperatures might be seasonal until established vegetation shades the ground.

60 MANAGING THE LOCAL CLIMATE

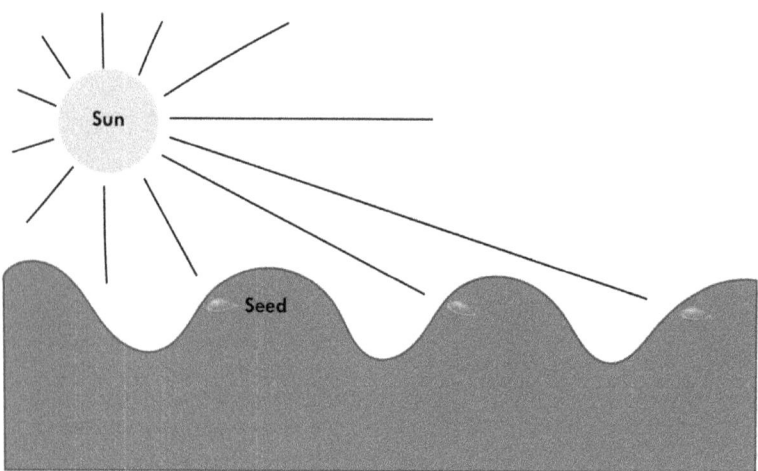

Figure 3.7 Illustration of seeds planted on the sunny side of the ridge to increase the received solar radiation when the sun's elevation is still low (winter and early spring)

A variation on the above is tie ridging, often used to increase water retention, especially in dry climates. Tie ridging implies obstructing the furrows repeatedly at a fixed interval. The ties allow water to rest in place and not to move laterally, and it has the additional benefit of preventing excessive breakage damage. Only the water stored between two ties will be released if a ridge breaks. Like in the case of normal ridging, this practice considerably augments soil moisture and crop production. Adeboye et al. (2017) found that tie ridging increased the soil water content in the lower 30–60 centimetres by 22 per cent. In another study, Araya and Stroosnijder (2010) showed that tied ridging increased the soil moisture in the root zone by more than 13 per cent compared with the control. This was linked to increased water retention and a reduction of seasonal runoff from 16–30 per cent to 5–9 per cent. Milkias et al. (2018) found a 134.59 per cent increase in soil moisture storage with tied ridging.

3.3 Management and conservation of vegetation

When the companies come, these swamps are drained, forests are felled, and we see that the lagoons and rivers begin to dry up and water retention is less. This affects the whole landscape, the rainfall, the winds, and the exposure to the sun

Justino Piaguaje, elder of the Siekopai indigenous people, Ecuador

The preservation of existing forests and additional well-managed regreening is central to managing local climates. There is much

> **Box 3.10** Protecting the rainforest in the Amazon, Ecuador
>
> There is enormous pressure on the Amazon area, from oil extraction and mining operations and the development of palm oil plantations. Particularly the conversion of rainforest into plantations changes the local climate significantly: the wind directions are altered, and the exposure to the sun increases in a way never experienced before. Moreover, the interventions set in motion a whole set of detrimental ecological changes – rivers become more sedimented, eddies that are breeding places for animal species disappear, and flood plains are covered in water, affecting nesting sites. The problem is not only the felling of the forests but also the indiscriminate machine-operated way in which this is done. The indigenous communities in the Amazon in Ecuador take many steps to protect the native forests: seeking legal protection, discouraging the new concessions, and rallying community action around the protection of the holy ceibo trees.

anecdotal evidence that where forests were cleared, local climates changed, particularly temperatures rising and rainfall dropping (Schwartz, 2013). Box 3.10 shows an example of how the indigenous communities in the Amazon put much effort into protecting their native forests to stop this from happening.

Vegetation is pivotal for many reasons. First, vegetation impacts the energy balance of a place as it affects how much heat is absorbed and reflected. A dense canopy forms an active surface above the crop zone, intercepting much solar radiation (Wilken, 1972). By absorbing and reflecting the incoming radiation before it reaches the soil, vegetation will hinder the increase of the ground temperature in the daytime and cooling at night, resulting in more moderate temperatures. Forest canopy closure can thus buffer local climatic biotic responses to global warming (De Frenne et al., 2013). Because of the plant's transpiration processes, the temperature on a vegetation canopy will be lower than that of bare soil (Stoutjesdijk and Barkman, 1992). Davis et al. (2019) found that under at least 50 per cent forest canopy cover, the maximum temperature was 5.3°C lower compared to areas without canopy cover. This study was conducted in the north-western United States.

The temperature effect extends beyond the forest area itself. Lejeune et al. (2018) established that deforestation in moderate regions contributed as much as 1°C to maximum temperature and contributed one-third to peak temperature days in places that lost more than 15 per cent of their forest cover. Crompton et al. (2021), based on research from South-east Asian coastal areas, found that deforestation increased land surface temperatures in the deforested area and the adjacent undisturbed areas. The effect would be noticeable as far as 6 kilometres away, with the temperature increase tapering off from 3.1°C at 1–2-kilometre distance to 0.75°C at a 6-kilometre distance.

Temperature increases were larger where deforestation was extensive and lower when clearing produced more fragmented landscapes in which non-forest and forest edges were intertwined. The recommendation is to have intact forests within 4 kilometres of farmland to offset the temperature rise due to global warming.

Second, vegetation can alter wind's speed and direction as it influences air temperature circulation at different layers (Ismangil et al., 2016). The trees' trunks, branches, and leaves reduce the wind speed by friction. However, when the paths open for airflow are narrowed, the wind speed increases and the rows of single trees then produce a 'funnel' effect (Geiger et al., 2003). When planting vegetation in a row, it is thus important to take the prevailing wind direction into account and ensure that the row is perpendicular. The protective effect of trees is a function of distance from the row of trees – the denser and higher the obstruction, the more significant the wind reduction immediately behind it. However, an obstruction should also not be too dense as this might result in turbulent eddies (see Section 3.5 for more information).

Third, forests are moisture sinks. A great moisture reservoir is stored in the forests, from the soil underneath the forest up to the treetops. The root of vegetation enhances moisture infiltration in the soil. The thick roots from trees result in large macropores that enable water to infiltrate deeper, even in otherwise impervious soils (Aubertin, 1971). Tree roots anchor the soil, retain water, and reduce erosion (Critchley et al., 2013). The mould in the forest floor acts as a moisture sponge. This vegetation litter improves soil structure and micro-organic life, determining soils' porosity and water holding capacity (Wiegant and Van Steenbergen, 2017). Next, the roots of trees can also serve as a hydraulic lift by bringing the moisture from the deeper soil layers to the dry surface layers (Zapater et al., 2011). Subsequently, vegetation canopy can retain moisture in the air, increasing air humidity levels.

Closely related to this, forests also contribute to rainfall reliability – though the geographical scale is debated. More vegetation allows more soil moisture, evapotranspiration, and relative humidity. A 10 per cent rise in relative humidity can lead to two-to-three times the amount of precipitation (Ellison et al., 2017). Increased soil moisture enhances evaporation and root moisture absorption and, consequently, transpiration through plants. Both processes contribute to precipitation not necessarily at an immediate local scale but surely at a regional scale. Air that passes over forests for 10 days typically produces at least twice as much rain as air that passes over sparse vegetation (Spracklen et al., 2012).

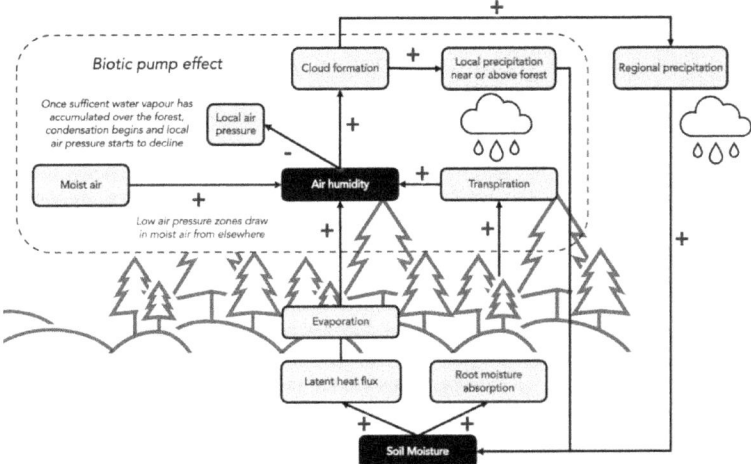

Figure 3.8 Overview of the hydrological cycle through evaporation and transpiration processes and the biotic pump effect. A '+' indicates one component having a positive (increasing) effect on the other component and a '−' indicates one component having a negative (decreasing) effect on the other component
Source: MetaMeta

Yet forests are also seen to attract rain. An important phenomenon explaining why it often rains locally above or in the close vicinity of forests is the 'biotic pump' effect (Makarieva and Gorshkovak, 2006; Ellison et al., 2012). Condensing vapour above forest areas creates low air pressure zones that draw in moist air from elsewhere (Figure 3.8).

This effect has been described both for evergreen as well as deciduous forests. In the latter, the effect occurs only in the summer, as in the winter, the forests are primarily leafless and dormant, and no vapour is released. Being evergreen thus helps trees to enhance winter precipitation, while deciduous species lack this feature (Makarieva and Gorshkovak, 2006). An anecdotal example of how a restored forest returned rainfall can be found in Madhya Pradesh, India (Box 3.11).

A study by Mekonen and Tesfahunegn (2011) on the Medogo watershed in northern Ethiopia stated that an increasing vegetation cover indeed improved its local climate. Respondents in this study explained that the hot and dry air that dominated the watershed had been replaced by moist and cooler air. Figure 3.9 depicts the difference between a bare plot and a plot with many trees. This figure shows that trees increase the evaporation fluxes compared to the sensible heat fluxes, resulting in a cooling effect. Using the sun's energy, individual

> **Box 3.11** Madhya Pradesh, India: restoring forest, restoring rainfall
>
> The hillsides near the villages of Manegaon and Dungariya in Madhya Pradesh's Sagar district once were a stretch of barren land. Now this area has turned into a lush, green 417-hectare forest with the help of the commons programmes of the Foundation for Ecological Security. The forest results from the loving care from people from the two villages in Deori Tehsil, who protected the young tree saplings for 20 years, watching them grow into adult trees that they are very proud of today. The roots of the trees were always there. By protecting them from interference (Photo 3.21), the forest could come alive again and regenerate on its own. The people work together with the forest as a living organism, being present when needed: for example, guiding newly arising streams and retaining the water for the plants and animals. In turn, the forest pays back generously. A regular rain cycle has now developed around the forest, solving the water shortage issue in the area. Within a 15-kilometre radius around the forest, the groundwater has slowly recharged, enabling the farmers to successfully grow a wide variety of crops (Jha, 2021).
>
>
>
> **Photo 3.21** Villagers checking on the forest in the evening
> *Credit:* Rajnish Mishra

trees can transpire hundreds of litres of water per day; this represents a cooling power equivalent to 70 kWh for every 100 litres of water transpired, which is enough to power two average household central air-conditioning units per day (Ellison et al., 2017).

It is argued that this climate regulating function of forests, in terms of temperature and rainfall control at local and regional scales, should be recognized as their principal contribution, with carbon storage and timber as co-benefits (Locatelli et al., 2016).

When trees are grown at a strategically chosen location, they can also provide other benefits, such as protecting cropland against harsh

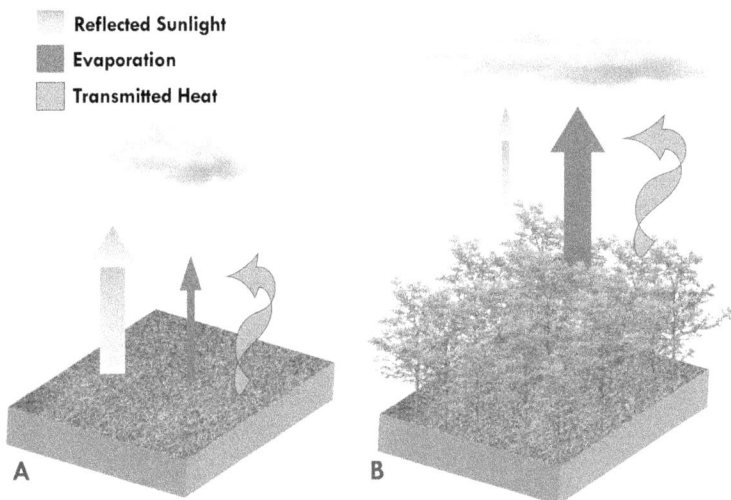

Figure 3.9 Examples of various biophysical factors in grassland or cropland (A) and forest (B). Because of cropland's higher reflectivity (albedo), it typically reflects more sunlight than the forest does, cooling surface air temperatures relatively more. In contrast, the forest often evaporates more water and transmits more heat to the atmosphere, cooling it locally compared to the cropland. More water vapour in the atmosphere can lead to a greater number and height of clouds as well as to increased convective rainfall
Source: Adapted from Jackson et al. (2008. © *IOP Publishing*. Reproduced with permission. All rights reserved)

winds, storms, or dust. In this light, roadside tree planting within traffic safety requirements is a particularly underutilized opportunity (Box 3.12). However, it should be noted that eddies might occur near the edges of tree patches, and a funnelling enhancing wind speed might be created. The location and density of trees as windbreaks is therefore critical. More information on trees as windbreaks is provided in Section 3.5.

While there is thus plenty of scientific and anecdotal evidence that tree cover influences the local climate, not all tree cover has an equal effect. One should be aware of several nuances when considering the regreening practices described further in this section. Different types of forests have other effects. Tree age and tree diversity are examples of this, which are discussed later on. Furthermore, the hydrological impacts of forests are also context dependent as a site can only influence water that is available in the vicinity (Sheil, 2018). Not taking this into account can lead to disappointing results. Afforestation thus requires consideration of environmental suitability (Ahrends et al., 2017).

Box 3.12 Roadside tree planting

Planting trees, shrubs, and grasses along the road is an often-overlooked option to create a productive asset and alleviate the harmful effects of roads on the local environment (Photo 3.22). Adverse effects include erosion, loss of fertile soils, gully formation that undermines road foundations, and heavy dust. Dust lifted by vehicles along unpaved roads directly affects the health of people and livestock living near the roads and crop production.

Photo 3.22 Row of metasequoia trees in Makino, Japan. The trees were planted to protect chestnut plantations

Road water harvesting and roadside plantations should therefore be combined, especially in low rainfall areas. Small diversion channels can be constructed to slowly divert surface flow from the roadside drainage system toward the tree seedlings. These diversion structures can be combined with small storage structures around the trees to retain this water for the tree. Smaller bushes and grasses can complement water harvesting by slowing down overland drainage flow (Van Steenbergen and Agujetas, 2018).

Several approaches to preserving and expanding vegetation are discussed in this section: forest preservation and management, farmer managed natural regeneration (FMNR), reforestation, agroforestry, and vegetation strips.

Forest preservation and management

The adverse effects of the complete loss or the fragmentation of old-growth forests in structurally simplified production forests can be avoided through forest preservation and management. In his book

The Hidden Life of Trees, Peter Wohlleben describes how trees should be seen as more than just a commodity and how selective logging and pruning, instead of cutting down the whole tree or entire forest patches, contributes to a species-appropriate manner of forest preservation (Wohlleben, 2016). Several other sustainable forest management practices are preventing the spread of pathogens and limiting illness of trees, preventing forest fires, and protecting the forest from illegal logging. Preserving an already existing forest through sustainable forest management practices allows reaping the more substantial thermal buffering ability benefits that old-growth forests have compared to young planted forests (Norris et al., 2011; Ellison et al., 2017; Sheil, 2018; Lin et al., 2020).

Farmer managed natural regeneration (FMNR)

FMNR is a set of practices used by farmers to encourage the growth of native trees on agricultural land. It is widely promoted in Africa as a cost-effective way of restoring degraded land that overcomes the challenge of low survival rates associated with planting new trees in arid and semi-arid areas. FMNR consists of thinning and pruning already existing stems in the field, which enhances their growth. Box 3.13 provides an example of FMNR in Tanzania.

Growing trees on agricultural land will, when a certain canopy coverage is reached, have climate regulating effects. The denser the canopy cover, the stronger the buffering effect of the local climate (Benítez et al., 2015). This is also true for tree height, as higher crowns generally provide more overstorey cover, resulting in a more substantial buffering effect (Bonan, 2016). When using FMNR to create a temperature buffering effect on a site, it is thus recommended to do this for approximately 50 trees per hectare. The other benefits of FMNR, such as increased infiltration capacity and enhanced soil organic matter levels, will be reaped earlier.

Reforestation

Increasing tree canopy cover through reforestation buffers temperatures and assists in moisture conservation. Reforestation and restoring degraded forest landscapes on an adequate scale may also reactivate the biotic pump and return the rainfall (Ellison et al., 2017). However, it should be noted that while a large area of planted forest will certainly influence the local water cycle, it is not yet clear how, when, and where this can replace the various properties and functions of a natural forest. Younger vegetation has distinct properties compared to more mature vegetation (Sheil, 2018).

Box 3.13 Dodoma, Tanzania: farmer managed natural revegetation

To regreen harsh degraded savannah areas, one of the most viable strategies is FMNR. Revegetation is based on dormant tree stumps that rest in the soils and sprout again after an occasional rainfall or flooding event. In FMNR, only beneficial indigenous species are protected. In the Dodoma region in Tanzania, Justdiggit and LEAD Foundation have applied the FMNR agroforestry technique, locally named *Kisiki hai*, to foster land restoration by increasing the number of valuable trees in the landscapes. *Kisiki hai* uses stumps already present in the field. The most vigorous stems of suitable stumps are pollarded; trees are regrown in one to two years (Photo 3.23). In many areas of sub-Saharan Africa, there are so many stumps that this can be considered a potential 'underground forest', ready to grow with only little cost and effort as each stump can produce from 10 up to 30 stems per year (Francis and Weston, 2015).

Photo 3.23 Trees grown with the *Kisiki hai* agroforestry technique after three years
Source: Justdiggit

The local climatic benefits of FMNR in Dodoma were investigated using field observations, drones, and satellites for nine study areas of approximately 50 hectares, spread throughout the region (Villani et al. 2020). This research showed that for areas with more than 7.2 per cent tree canopy cover, there is a significant effect with a reduced land surface temperature of up to 1.32°C.

Reforestation consists of several steps: First, seeds need to be collected and stored. Second, tree nurseries need to be established. Third, once grown, these small trees need to be out planted. Fourth, after plantation in the field, post-planting treatment and monitoring are required to increase post-planting survival rates. Increasing the sapling survival rates can be done by increasing water availability through microbasins (see Section 3.1), planting young trees near adult remnant trees, planting multiple sapling species (Sprenkle-Hyppolite et al., 2016), and protecting the young trees from harsh winds. Once the trees are established and the roots have grown deeper, the trees themselves can serve as a windbreak (Sun and Dickinson, 1994). The choice of tree species in reforestation practices is also significant for the effect the trees will have on the below-canopy climate. For instance, one could make deliberate choices on the shade casting ability, rotation lengths, and presence of a shrub layer (De Frenne et al., 2021).

Agroforestry

In an agroforestry system, annual crops and perennial plants are cultivated in the same field with different spatial and temporal arrangements. There are multiple examples of agroforestry worldwide, and farmers may arrange trees according to their preference and farming practice. A particular agroforestry layout is alley farming: cultivating annual crops in between tree rows (Photo 3.24).

Photo 3.24 Example of an alley cropping system in Hararghe, Ethiopia. Here maize is grown in the alley between chat rows
Credit: Francesco Sambalino

Another example is multi-strata agroforestry consisting of multiple trees, mimicking a forest, with crops growing in between. Here it is important to note that the temperature buffering effect of forests is reduced near forest edges, with the edge effect extending up to 20 metres inside the forest (Ewers and Banks-Leite, 2013). This implies creating a buffer zone forest around the crop growing area. Sida et al. (2018) found a decrease in midday air temperature of about 6°C under the trees in an agroforestry system compared to open fields. This study was conducted in the Central Rift Valley, Ethiopia.

In agroforestry, the choice is made to dedicate land to tree crops rather than field crops, but this is not a zero-sum decision (Gardner et al., 1999). Apart from the direct return from the agroforestry systems, yields are increased on the adjacent land due to reduced wind erosion, improved local climate, and increased soil moisture (Geiger et al., 2003; Jones, 2014). Lin (2007) also found that using an agroforestry system is economically feasible to protect crop plants from extremes in local climate and soil moisture.

The trees also impact the net radiation received by the crops. When planted in rows, the shading effect is much less on north–south oriented trees than on east–west oriented rows (Geiger et al., 2003). In areas with high interception losses, it might be beneficial to use smaller-leafed deciduous tree species that reduce the interception. Another solution can be management practices such as thinning and tree pruning.

During the night, isolated small woodlands can be warm islands in the sea of cool grasslands and fields around. Even a single large tree has a warmer microclimate at night than the surroundings. Crops thus can receive some long-wave radiation from the trees planted on farms. However, this influence diminishes with decreasing distance from the tree row (Stoutjesdijk and Barkman, 1992). To ensure the temperature buffering effect from trees, it is thus recommended to add multiple small woodlands over a landscape or multiple single large trees or rows of trees on a farm.

Vegetation strips

One way to create vegetation strips is by growing grass edges across farmland or on its boundaries (Mati, 2006) (Photo 3.25). When planted following the farm contour lines, they help counter soil erosion processes. Grass strips form a (semi-)permeable barrier that slows down the water flow and traps loose soil particles (Critchley et al., 2013).

Because of this process, more water infiltrates the ground and causes an increase in soil moisture (Mati, 2006). Dass et al. (2011)

Photo 3.25 Grass lines
Credit: Mathias Gurtner

found that plots with a vegetative barrier of *Sambuta* in Orissa, India, had a higher moisture content (17.1 per cent) compared to the control plot (13.5 per cent). The higher moisture content was due to greater conservation and infiltration of rain and runoff water into the soil. Patil et al. (1995) observed 16 per cent higher soil moisture in the sorghum cropped plots when provided with a vetiver barrier than in the control plot.

Grass strips influence the local climate and may be compared to other revegetation measures that involve trees and shrubs, albeit on a smaller scale. Wind speed, for example, is influenced to a minor degree because this function is directly correlated to vegetation height.

Hedgerows are another form of vegetation strip and majorly control many landscape fluxes such as wind speed, evapotranspiration, soil erosion, and species movement (Forman and Baudry, 1984). More information on the effect of hedgerows on the local climate through reduced wind speed is provided in Section 3.5.

The management and conservation of vegetation is thus of high importance for a well-buffered local climate. All techniques described in this section – forest preservation, revegetation, FMNR, agroforestry, vegetation strips – can be combined to create the best local climate optimization (Figure 3.10).

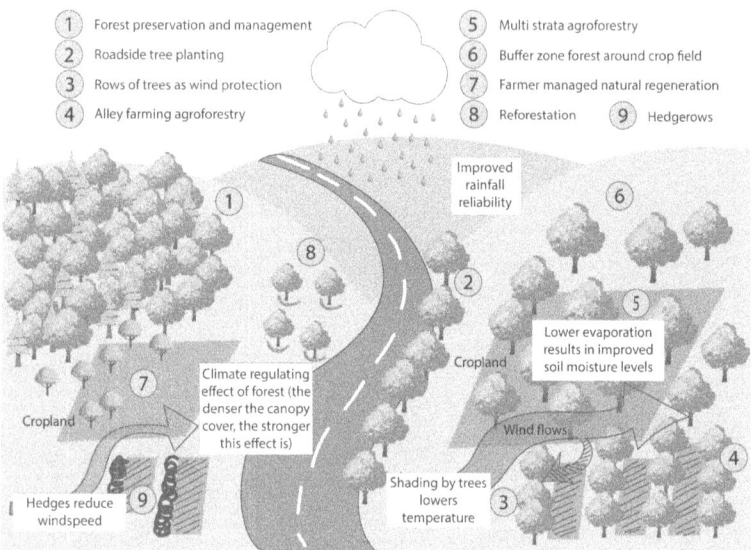

Figure 3.10 An overview of several vegetation conservation and management techniques for local climate improvement implemented in a landscape
Source: MetaMeta

3.4 Solar radiation management

This section elaborates on solar radiation management. Solar radiation, particularly the net radiation that contributes to the energy balance, is of high importance in determining the local climate. Solar radiation of the correct intensity, quality, and duration is also essential for normal plant development as it provides plants with the energy to convert carbon dioxide into sugars. The amount of solar radiation that is received by a plot can be managed by row orientation, row spacing, and shading.

Row orientation

The latitude of the farm can inform farmers' choices to determine crops' row orientation when they want to influence the amount of solar radiation that is received by crops and soil and control micro-climatic conditions (Wilken, 1972). When the sunlight hits the rows at a right angle, crop production may increase due to the increased light interception by crops and the shading of weeds in the inter-row spaces (Borger et al., 2016). When the orientation is north–south in the northern temperate hemisphere, crops intercept more solar radiation than the soil, reducing soil temperature. In India, higher relative humidity was recorded in the north–south row direction (Sandhu and

Dhaliwal, 2016). In the case of an east–west direction in the northern hemisphere, the soil intercepts more solar radiation, leading to an increase in soil temperature (Yildiz and Rattan, 1996), and increased water evaporation (Gegner et al., 2008).

Row orientation can also reduce wind erosion when the rows are perpendicular to the prevailing wind direction (Funk and Engel, 2015); see Section 3.5. Directing the rows of vegetation or crops in a suitable orientation can thus benefit the farm's climate in terms of soil and water conservation and buffered temperatures.

Row spacing

Changing crop row spacing can maximize or minimize light penetration and trap short- and long-wave radiation. A study by Yang et al. (2008) found that with increased row spacing, the relative humidity decreased, and the canopy temperature increased. They also stated that crop yield could be improved by reduced row spacing. Stickler and Laude (1960) found that the soil temperature, light intensity, and evaporation from the soil surface decreased with decreasing row spacing. So, the closer the rows, the lower the soil temperature and evaporation rates, resulting in improved soil moisture levels (Dhaliwal et al., 2019). A study from Sandhu and Dhaliwal (2016) on row spacing and relative humidity profiles within crops found that the smaller the row spacing, the higher the air humidity on crop level. In 15-centimetre row spacing, almost 4 per cent more relative humidity was recorded than for 30-centimetre row spacing. Any yield advantage to growing crops in narrow rows may result from establishing a more uniform root and leaf distribution that aids in exploiting soil water and light resources and reducing soil temperatures and evaporation compared with crop production in conventional rows (Sharratt and McWilliams, 2005). However, Aubertin and Peters (1961) showed that when narrow crop row spacing was conducted under moisture shortage conditions, radiant energy's efficient capture may lead to significantly increased transpiration and, hence, to severe wilting.

Row spacing has the potential to change climatic conditions under the crop canopy. However, a balance needs to be achieved to avoid overcrowding and the risk of severe competition between plants.

Shading

Although photosynthesis rates typically increase with higher irradiance, too high irradiances can damage the photosynthetic system, mainly when other stresses such as extreme temperature or water stress occur

> **Box 3.14** Using shade nets for leafy vegetables in central Spain
>
> There is a ready market for lettuce and leafy vegetables in Madrid in central Spain. Unfortunately, intense heat in the summer period stresses these vegetables. To sustain production in this period, ways to lower the immediate temperature around the crops had to be found. To resolve this, shade nets are used above the crops (Photo 3.26). These nets filter 50 per cent of the solar radiation, which causes a reduction in temperature since crops and the soil receive less sunlight. This lower temperature prevents the lettuce from flowering too early, which would make the crop unsellable. This use of shade net is an example of how, despite extreme summer heat conditions, the immediate weather environment of the crops can be managed.
>
>
>
> **Photo 3.26** Shade nets on a small market garden in central Spain
> *Credit:* Francesco Sambalino

(Jones, 2014). In these cases, shading can be helpful to moderate the light intensity. Shading affects the energy available for heating and evaporation, and it may lower soil temperatures and reduce water evapotranspiration, maintaining higher soil moisture levels. The lower evaporative demand under shade allows plants to increase stomatal conductance and CO_2 assimilation (Mahmood et al., 2018). Box 3.14 displays an example of the use of shade nets to grow leafy vegetables in the hot climate of central Spain.

Shade nets can also protect against frosts and chilling effects by capturing the long-wave radiation that would have escaped easily in the open air (Stigter, 1984b). Box 3.15 displays an impressive example from Tajikistan where citrus plants are grown in extreme cold weather conditions.

Box 3.15 Growing citrus in extreme weather conditions in Tajikistan

Citrus fruits are used extensively in tea preparation in Central Asia. But citrus plants can usually only be grown in tropical and subtropical climates as they cannot cope with frost. Therefore, for an extended period, they were imported from warmer areas abroad.

However, the region wanted to become self-sufficient in citrus production. Considerable effort was put into researching how the fruits could be grown locally despite the harsh weather conditions. Part of the solution, next to breeding cold-resistant varieties, was to create favourable microclimates for the citrus plants to grow in. These microclimates were created by digging trenches to protect the plants from harsh weather and be nourished by the soil heat. Trenches were located on level ground or light slopes, oriented from east to west for optimal sunlight during the winter months. Farmers could place shade screens or shading plants between the trenches to protect the citrus plants from overheating in summer. When winter came, farmers covered the trenches with thin wooden boards or straw mats, maintaining the soil heat in the trench while keeping snow and precipitation out. These efforts extended the area of citrus cultivation along the region westward of the Black Sea coast, Uzbekistan, and the southern districts of Ukraine and Moldova (De Decker, 2020).

Photo 3.27 Example of shading in a tree nursery in Amhara region, Ethiopia
Credit: Jean Marc Pace Ricci

Shading also increases the absolute air humidity, decreasing the evaporative demand. Mahmood et al. (2018) found that shading screens increased relative humidity by 2–21 per cent, reduced air temperature by 2.3–2.5°C, and decreased evapotranspiration by 17.4–50 per cent.

Several studies also showed that shading screens could provide physical protection against hail, wind, bird, and insect-transmitted virus diseases (Shahak, 2008). Tanny and Cohen (2003) found that a shade net reduced the wind speed by about 40 per cent compared to an unshaded site.

It should be considered that too little light intensity can decrease crop production if the crop is not getting enough light for the photosynthesis process (Rosenberg et al., 1983). Shading is thus helpful to facilitate the growth of shade-tolerant crops or where light intensity is excessive (Photo 3.27).

3.5 Wind management

This section elaborates on managing the wind on a site and shows the example of the effect of windbreaks on the local climate and crop production.

Windbreaks

A windbreak is a barrier that influences the wind direction and speed. The length, width, height, and material may vary considerably. Some examples of windbreaks are trees (Photo 3.28) with and without undergrowth, hedges and rows of bushes, tall plants (e.g. sunflowers), straw matting, wire netting with small mesh, stone walls, board fences, and fences of woven reeds (Geiger et al., 2003). Shelterbelts (trees functioning as a living barrier surrounding a plot) also function as windbreaks (Photo 3.29). Box 3.16 shows an example of how hedges and trees serve as a windbreak on a Dutch farm.

Windbreaks vary in effectiveness, depending upon their height, porosity, and length. A barrier should allow just enough air to penetrate to prevent eddying on the lee side but still reduce the wind speed enough (Figure 3.11) (Allaby, 2015). A porosity of 50 per cent results in a downwind influence of 20–25 times the height of the barrier. Thus, the higher the windbreak, the greater the distance of its downwind and upwind influence. The longer the windbreak, the more constant its influence. If a barrier is too short, jetting effects may increase rather than reduce wind speed (Caborn, 1957). Also, the windbreak should be placed perpendicular to the prevailing wind direction; otherwise, it might funnel the wind and increase wind speed (Allaby, 2015). Windbreaks can also be placed at crop level.

When using vegetation as windbreaks, its effectiveness depends on the vegetation used. When a windbreak consists of deciduous trees, the leaf index changes throughout the seasons, resulting in less windbreak effect in winter due to low leaf index. This can be solved by, for example, planting multiple rows or using a combination of evergreen and deciduous species.

IMPROVING THE LOCAL CLIMATE 77

Photo 3.28 Tall green wind shield edges and kiwi cultivation, shot in bright late spring light near Wakamarama, Bay of Plenty, North Island, New Zealand

Photo 3.29 Aerial view of landscape plots surrounded by shelterbelts serving as windbreak

One benefit of windbreaks is reduced soil erosion (Jensen, 1954). Winds can detach small soil particles from the soil surface, and this detachment can result in the loss of topsoil, which reduces the farmable soil depth and soil fertility. Reduced soil fertility can result in a loss of vegetation potential and soil dehydration (Foken, 2008), which negatively affects the local climate, as discussed in the earlier sections. Windbreaks can prevent this by slowing down the wind

> **Box 3.16** Silvopasture against weather extremes in Stoutenburg, the Netherlands
>
> The Netherlands is classified as a moderate maritime climate, characterized by mild winters and cool summers with precipitation all year round. Climatic extremes, however, are getting more and more common, affecting livestock farming. During heatwaves in the summer, farmers have to keep the cows indoors. In winter, harsh cold winds can blow against the stables, threatening the cows' health. On the Van Zandbrink's biological livestock farm in Stoutenburg called 'Boerderij Tussen de Hagen' (translated: farm between hedgerows), measures were taken to buffer against these climatic challenges.
>
> One of the solutions has been silvopasture, that is, integrating tree cultivation and animal production. Rows of fruit trees have been planted where the calves can graze in the shade. To create a sheltered local climate, they surrounded the plots with hedges and trees as a windbreak. To avoid the harsh cold wind blowing on the stable in the winter, a hedgerow was grown in front of the building (Photo 3.30). This hedgerow breaks the wind and provides shelter for the cows to go outside the stable, even on winter days, by creating a better climate on the farm.
>
>
>
> **Photo 3.30** Hedgerow in front of the cow stable to break down the harsh south-western wind, which creates a more conducive climate for the cows

speed. Another effect of windbreaks is the mechanical protection of the crops, as winds can cause physical stress and damage to crops (Rosenberg et al., 1983).

Windbreaks reduce the transport and mixing of air. The lowered wind speed decreases the interaction between the air layers above the surface. This results in steeper vertical air temperature profiles in sheltered fields than open fields without a windbreak. As a result of this lower air mixing activity, the near-surface air temperatures in the sheltered zone are higher during the day. When the sensible heat flux is typically negative at night, this effect leads to cooler near-surface air

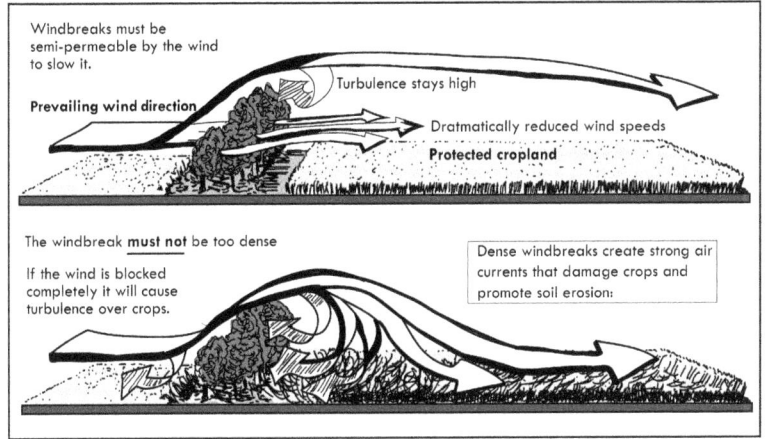

Figure 3.11 Visual representation of the effect of the density of a windbreak on the wind flow
Source: Adapted from Tengnäs, 1994

temperatures in the sheltered area (Cleugh, 1998). Outside the sheltered zone, the surface radiative heat losses are replenished faster due to the wind, resulting in higher air temperatures during the night (Oke, 1995). Windbreaks thus result in higher daytime and lower night-time temperatures due to less mixing of the inversion layer (Rosenberg et al., 1983). Campi et al. (2009) found that the air temperature increased by up to 9.4 per cent when the wind speed was reduced by 65 per cent. This study was conducted in a Mediterranean environment.

The higher air temperatures in the shelter may extend to the soil (Geiger et al., 2003). Higher soil temperature will result in more rapid root respiration and organic matter decomposition and cause a more significant release of CO_2 from the soil (Rosenberg et al., 1983). The higher soil temperatures may also result in rapid seed germination (Rosenberg et al., 1983), but if in excess, may lead to poor seed germination (Covell et al., 1986).

The lowered wind speed also results in a higher daytime relative humidity because of less vertical mixing (Geiger et al., 2003). Marshall (1967) found an increase of 3 per cent near the wind barrier. Dew is also enhanced in sheltered areas due to greater humidity and colder night-time temperatures (Oke, 1995). This dew can be a potential moisture source. However, it should be noted that this can also lead to increased fungal diseases (Rosenberg et al., 1983).

Wind speed reduction also lowers evaporation processes, improving water use efficiency (Cleugh, 1998). The higher the wind speed, the higher the evapotranspiration rate, explained in Chapter 2. A too

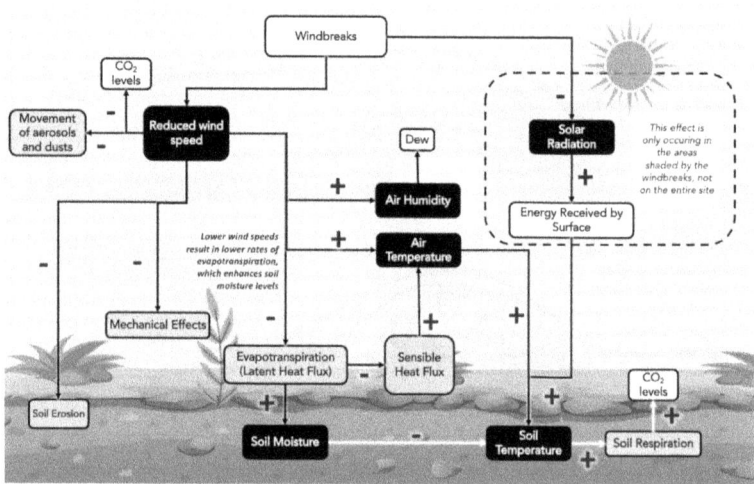

Figure 3.12 Overview of the daytime effect of windbreaks on wind direction and speed and the soil temperature, air temperature, air humidity, and soil moisture interactions. A '+' indicates one component having a positive (increasing) effect on the other components and a '−' indicates one component having a negative (decreasing) effect on the other components
Source: MetaMeta

high evapotranspiration rate can impair plant water status, negatively affecting agricultural productivity. In cases of low soil moisture, windbreaks can thus be effective measures in lowering the evaporation rate and thereby maintaining higher soil moisture levels.

Many studies show that windbreaks increase crop yield in nearly all cases where it has been tried in a wide range of climatic regimes (Grace, 1977; Geiger et al., 2003). Despite the wind being essential in replenishing the CO_2 levels needed for photosynthetic activity, Brown and Rosenberg (1972) found that the daytime reductions in CO_2 in a shelter were not significant and certainly would not affect photosynthetic activity. An overview of the daytime effect of windbreaks on the local climate is given in Figure 3.12.

3.6 Finding the best local climate management practices for a specific site

The previous sections of this chapter showed a range of practices that farmers can use to manage the local climate. The most suitable approach for a particular site depends on the improvement that is desired and the characteristics of that specific landscape. For example, some interventions will not be possible in an area with a very steep slope or without access to stones. Table 3.1 provides an overview of

Table 3.1 Guidance tool: expected local climatic impacts, the magnitude of impact, and landscape characteristic requirements for each practice

Local climate management practices	Expected local climatic impacts	Landscape characteristic requirements for implementation		
		Rainfall conditions	Slope	Soil conditions
Management of water and moisture				
In situ water infiltration (e.g. bunds, terracing, trenches, and swales)	• Increased in situ soil moisture levels (e.g. 5–10% increase was found in Ethiopia after implementing soil bunds (1), 16% increase was found in the Himalayan foothills after implementing trenches (2), 16% increase was found in Pakistan as a result of terracing (3)). • Terracing also optimizes the use of solar radiation and can provide shading for internal crop rows leading to a temperature reduction (e.g. 2°C lower in the morning (4)). • Improved water infiltration results in better buffered land surface temperatures (e.g. a 1.74°C land surface temperature reduction in Tigray, Ethiopia (5)). • Buffered surface temperatures indirectly result in better-buffered air temperatures in the area. **The magnitude of impact:** • The more consistently these practices are implemented throughout a landscape, the stronger the climate regulating effect.	Bunds: In semi-arid and arid regions and medium rainfall areas (6). Trenches: Suited to rainfall conditions below 900 mm annual rainfall (8). Terraces: Arid to wet regions (8).	Bunds: Maximum of 20%. Above 15% slope, it is better to use stone bunds (7). Trenches: Suited to slopes below 35% (8). Terraces: On average, 12–58% (6).	Bunds: Suitable for most soils. However, it should be strengthened with stones on heavy black cotton and sandy soils (7). Trenches: Well-draining soils (6). Not suitable on shallow soils (8). Terraces: Deep soils (6). Not suitable for sandy or stony soils (8).

(Continued)

82 MANAGING THE LOCAL CLIMATE

Table 3.1 Continued

Local climate management practices	Expected local climatic impacts	Landscape characteristic requirements for implementation		
		Rainfall conditions	Slope	Soil conditions
Small storage pockets (e.g. semi-circular bunds, zaï pits, eyebrow basins)	• Increased in situ soil moisture levels (e.g. an increase from 0.085% to 1.62% was found in a desert environment after implementing semi-circular bunds (9), a surface runoff reduction of 19% due to eyebrow terraces (12). • Better-buffered temperatures. • Increased air humidity levels. • Improved vegetation and crop growth. **The magnitude of impact:** • The more consistently these practices are implemented throughout a landscape, the stronger the climate regulating effect.	Eyebrow basins: Annual rainfall below 1,400 mm (8). Semi-circular bunds: 200–750 mm of annual rainfall (arid to wet areas) (10). Zaï pits: Arid to wet regions (11).	Eyebrow basins: Applicable in steep and degraded hillsides (6). Semi-circular bunds: Suited to slopes up to 15% (11). Zaï pits: Slopes up to 15% (11).	Eyebrow basins: Suited for medium well-drained deep soils, with a minimum depth of 25 cm (8). Not suited for vertisol. Semi-circular bunds: Moderately to deep soils required (8). Zaï pits: Suitable for all soils.
Barriers to concentrate runoff or to divert floodwater from local rivers and streams (e.g. check dams, spate irrigation, or floodplain farming systems)	• Higher soil moisture levels (e.g. check dams were found to improve flow volumes, supported by increased groundwater levels, by 28% (13)). • Improved temperature buffering capacity due to water diversion (A maximum surface temperature difference of 12°C between exposed sediments and an aquatic habitat was found in Alpine river floodplains (14)).	Check dams: Suited to all rainfall conditions (8).	Check dams: 1–35% (8).	Check dams: Suitable for all soils (7).

(Continued)

Table 3.1 Continued

Local climate management practices	Expected local climatic impacts	Landscape characteristic requirements for implementation		
		Rainfall conditions	Slope	Soil conditions
	• Buffered surface temperatures indirectly result in better-buffered air temperatures in the area. • Irrigating an area leads to cooler land surface temperatures by as much as 20°C (15). • Irrigated areas also enhance rainfall in areas downwind (16). **The magnitude of impact:** • The cooling effect of irrigated land can be extended to areas 5–40 kilometres from the irrigated region, depending on the wind speed (15).			
Water reservoirs (e.g. ponds, road water harvesting)	• Buffering effect on the temperature in the surrounding area (e.g. ponds can result in temperature differences of 4–5°C (17)). • Water bodies also increase air humidity around them, causing fog and dew. **The magnitude of impact:** • The buffering capacity of ponds can range up to 200 metres around the water bodies (17). • Deeper water bodies, such as lakes, were found to have a protective temperature buffer zone up to 5 kilometres (15).	Suited to all rainfall conditions.	Suitable for all slopes.	Suitable for all soils.

(Continued)

84 MANAGING THE LOCAL CLIMATE

Table 3.1 Continued

Local climate management practices	Expected local climatic impacts	Landscape characteristic requirements for implementation		
		Rainfall conditions	Slope	Soil conditions
	Adjusted agronomic practice			
Soil management (e.g. applying organic fertilizer or using green manure)	• Improved in situ soil moisture levels (e.g. a 55% increase was found after the application of organic fertilizer in combination with soil bunds (18)). • Better-buffered soil temperatures on the plot (e.g. a 2°C reduction and reduced day–night difference were found after biochar application (19)). • Buffered surface temperatures indirectly result in better-buffered air temperatures on the plot. • Dark soil amendments (e.g. biochar) lower the soil surface albedo, resulting in higher absorption of solar radiation. Pale materials (e.g. kaolin or ash) increase soil surface albedo and decrease soil temperatures by reflecting solar radiation. **The magnitude of impact:** • The larger the area these soil management practices are applied to, the more significant its effect on the local climate.	Suited to all rainfall conditions.	When farming operation is possible, soil management is possible.	Suitable for all soils.

(Continued)

Table 3.1 Continued

Local climate management practices	Expected local climatic impacts	Landscape characteristic requirements for implementation		
		Rainfall conditions	Slope	Soil conditions
Tillage (e.g. conservation tillage or no-tillage)	• Improved in situ soil moisture levels (e.g. an increase of 15% was found in Kentucky (20)). • Better-buffered soil temperatures on the plot (e.g. differences up to 1.5°C were found in China (21)). • Buffered surface temperatures indirectly result in better-buffered air temperatures on the plot. **The magnitude of impact:** • The larger the area these agronomic practices are applied to, the more significant its effect on the local climate.	Suited to all rainfall conditions.	When farming operation is possible, tillage is possible.	Deep soil (8).
Mulching (e.g. applying crop residues, straw, or plastic sheets)	• Improved soil moisture levels (e.g. a 4.2% increase was found in Ludhiana, India (22), a 17% increase was found in Nigeria (23)). • Non-hygroscopic mulch layers of a proper thickness can provide small amounts of dew to plant roots. • Improved soil temperature buffering capacity (e.g. 25.1–27.2°C fluctuations compared to 26.9–34.0°C (24)).	Suited to all rainfall conditions.	When farming operation is possible, soil management is possible.	Suitable for all soils.

(Continued)

Table 3.1 Continued

Local climate management practices	Expected local climatic impacts	Landscape characteristic requirements for implementation		
		Rainfall conditions	Slope	Soil conditions
	• Buffered surface temperatures indirectly result in better-buffered air temperatures on the plot. • Plastic sheet mulching increases a soil's temperature, assisting in frost protection (25). **The magnitude of impact:** • The larger the area mulching practices are applied to, the more significant its effect on the local climate.			
Intercropping	• Better-buffered in situ soil temperatures (e.g. 18.1–23.7°C compared to 22.6–28°C for a farm in Kenya (26)). • Improves soil moisture levels (e.g. a maximum of 27% compared to 18% in Kenya (26)). • Buffered surface temperatures indirectly result in better-buffered air temperatures on the plot. **The magnitude of impact:** • The larger the area intercropping practices are applied to, the more significant its effect on the local climate.	Suited to all rainfall conditions.	When farming operation is possible, intercropping is possible.	Suitable for all soils.

(Continued)

Table 3.1 Continued

Local climate management practices	Expected local climatic impacts	Landscape characteristic requirements for implementation		
		Rainfall conditions	Slope	Soil conditions
Furrowing and tie ridging	• Increased soil moisture levels (e.g. an increase of 121.87% was found for furrowing, and 134.59% for tie ridging in a field in Ethiopia (27), a 22% increase was found in Nigeria (23)). • This practice also impacts the amount of solar radiation reaching the ground, and better use of available incoming radiation can be accomplished. **The magnitude of impact:** • The larger the area these agronomic practices are applied to, the more significant its effect on the local climate.	Suited to all rainfall conditions.	Gentle slopes.	It may be difficult on heavy and compacted soil (10). Tie ridging is not suitable for sandy soils or poor infiltration soils (7).
Management and conservation of vegetation				
Forest and vegetation preservation and management	• More moderate soil and air temperatures, both in the forest as well as for the surrounding areas (e.g. a reduction of 5.3°C maximum temperature under at least 50% forest canopy cover was found in the north-western United States (28)). • Forests serve as moisture sinks. • The vegetation canopy can retain moisture in the air, increasing air humidity levels.	Suited to all rainfall conditions.	Suitable for all slopes.	Suitable for all soils.

(Continued)

Table 3.1 Continued

Local climate management practices	Expected local climatic impacts	Landscape characteristic requirements for implementation		
		Rainfall conditions	Slope	Soil conditions
	• Increase of both local and regional rainfall reliability. **The magnitude of impact:** • In South-east Asian coastal areas, it was found that the forest temperature buffering effect was noticeable as far as 6 kilometres away (29)).			
Reforestation (e.g. planting and FMNR)	• Improved temperature buffering effect once a certain canopy cover is established (e.g. 10% canopy cover resulted in a 1.32°C land surface temperature reduction in Tanzania (30)). • Buffered surface temperatures also indirectly result in better-buffered air temperatures in the area. • Improved moisture conservation capacity and increased air humidity levels. • When water is available in the vicinity, reforestation may reactivate the biotic pump and return rainfall (31). **The magnitude of impact:** • The denser the reforested canopy cover, the stronger the buffering effect of the microclimate (32). • The higher the trees, the stronger the buffering effect on the microclimate (33).	Suited to all rainfall conditions.	Suitable for all slopes.	Suitable for all soils.

(Continued)

Table 3.1 Continued

Local climate management practices	Expected local climatic impacts	Landscape characteristic requirements for implementation		
		Rainfall conditions	Slope	Soil conditions
Agroforestry (e.g. alley farming, multi-strata farming)	• Better-buffered temperatures (both cooling effect as well as frost protection) (e.g. a decrease of a midday air temperature of 6°C under trees was found in Ethiopia (34)). • Improved soil moisture conditions. • Wind speed reduction (when rows of trees are placed perpendicular to the prevailing wind). **The magnitude of impact:** • The larger the agroforestry system, the stronger its buffering effect. It is advised to design a buffering zone around the crop fields as the buffering effect reduces near the edges (35). • The more rows of trees or scattered woodlands over a plot, the stronger the temperature buffering effect.	Suited to all rainfall conditions.	When farming operation is possible, agroforestry is possible.	Suitable for all soils.
Vegetation strips (e.g. grass strips or hedgerows)	• Improved in situ moisture conservation (e.g. moisture content of 17.1% was found on a plot with vegetative barrier compared to 13.5% on a bare plot in India (36), a 16% increase was seen with vetiver grass (37)). • Better-buffered temperatures.	Suited to all rainfall conditions.	Grass strips: Gentle slopes up to 8% (7) or below 15% (8).	Suitable for all soils (7).

(Continued)

Table 3.1 Continued

Local climate management practices	Expected local climatic impacts	Landscape characteristic requirements for implementation		
		Rainfall conditions	Slope	Soil conditions
	• Lowered wind speed (when placed perpendicular to the prevailing wind). **The magnitude of impact:** • The longer and denser (not too dense) a vegetation strip, the more constant its wind reduction effect is.			
	Solar radiation management			
Row orientation	• Optimal solar interception for crops. • Less moisture evaporation from the soil improves in situ soil moisture levels. • Better-buffered temperatures. • Reduced wind erosion when placed perpendicular to prevailing wind. **The magnitude of impact:** • The larger the area this practice is applied to, the more significant its effect on the local climate.	When farming operation is possible, altering row orientation is possible.	When planting operation is possible, changing the row orientation is possible.	Suitable for all soils.
Smaller row spacing	• The in situ effects are conserved soil moisture, increased humidity, and buffered temperatures (22). • The smaller the row spacing, the higher the air humidity at crop level (e.g. a 4% higher relative air humidity was recorded in India (38)). **The magnitude of impact:** • Row spacing alters the microclimatic conditions below the crop canopy.	When farming operation is possible, altering row spacing is possible.	When planting operation is possible, changing the row orientation is possible.	Suitable for all soils.

(Continued)

Table 3.1 Continued

Local climate management practices	Expected local climatic impacts	Landscape characteristic requirements for implementation		
		Rainfall conditions	Slope	Soil conditions
	• The larger the area this practice is applied to, the more significant its effect on the local climate.			
Shading	• Less evaporation results in higher soil moisture levels (e.g. evapotranspiration was found to decrease by 17.4–50% (39). • Increased air humidity levels (e.g. increases of 2–21% were found (39)). • Better-buffered temperatures (e.g. a decrease of a midday air temperature of 6°C under trees was found in Ethiopia (34), a reduction of 2.5°C was found under shade screens (39)). • Depending on the shading material, shading can also reduce the wind speed. **The magnitude of impact:** • Shading directly influences the microclimate of the shaded area (e.g. the area under shade nets or trees). • In the case of shading by trees: the denser the canopy cover, the stronger the buffering effect of the microclimate (20). • There can be a cumulative effect of series of hedgerows and windbreaks on soil moisture levels and buffered air temperatures.	Suited to all rainfall conditions.	When farming operation is possible, shading is possible.	Suitable for all soils.

(Continued)

Table 3.1 Continued

Local climate management practices	Expected local climatic impacts	Landscape characteristic requirements for implementation		
		Rainfall conditions	Slope	Soil conditions
	Wind management			
Windbreaks	• Lower wind speeds (when placed perpendicular to the prevailing wind direction). • Reduced levels of soil erosion. • Higher temperatures (e.g. an increase of 9.4% in air temperature was found in the Mediterranean (40)). • Higher air humidity (e.g. an increase of 3% (41)). • Lower evaporation and improved soil moisture levels. **The magnitude of impact:** • A windbreak with a porosity of 50% results in a downwind influence of 20–25 times the height of the windbreak (42). • The longer and higher a windbreak, the more constant its influence; jetting effects may increase if a windbreak is too short (43).	Suited to all rainfall conditions.	Depending on the material used.	Suitable for all soils.

(1) (Nyssen et al., 2010)
(2) (Kaushal et al., 2021)
(3) (Rashid et al., 2016)
(4) (Tucci et al., 2019)
(5) (Castelli et al., 2018)
(6) (Ministry Agriculture Ethiopia, 2005)
(7) (Van Steenbergen et al., 2011)
(8) (HELVETAS Swiss Intercooperation et al., 2015)
(9) (Ahmed and Al-Dousari, 2014)
(10) (Critchley et al., 2013)
(11) (Nature^Squared et al., n.d.)
(12) (Hussain and Irfan, 2012)
(13) (Norman et al., 2015)
(14) (Tonolla et al., 2010)
(15) (Lawston et al., 2020)
(16) (Alter et al., 2015)
(17) (Nedjai et al., 2018)
(18) (Kuzucu, 2019)
(19) (Feng et al., 2021)
(20) (Blevins et al., 1971)
(21) (Shen et al., 2018)
(22) (Dhaliwal et al., 2019)
(23) (Adeboye et al., 2017)
(24) (Tayade et al., 2016)
(25) (Moreno and Moreno, 2008)
(26) (Nyawade et al., 2019)
(27) (Milkias et al., 2018)
(28) (Davis et al., 2019)
(29) (Crompton et al., 2021)
(30) (Villani et al., 2021)
(31) (Ellison et al., 2017)
(32) (Benítez et al., 2015)
(33) (Bonan, 2016)
(34) (Sida et al., 2018)
(35) (Ewers and Banks-Leite, 2013)
(36) (Dass et al., 2011)
(37) (Patil et al., 1995)
(38) (Sandhu and Dhaliwal, 2016)
(39) (Mahmood et al., 2018)
(40) (Campi et al., 2009)
(41) (Marshall, 1967)
(42) (Caborn, 1957)
(43) (Allaby, 2015)

Box 3.17 Transformed farms in Makueni County, Kenya

In Kenya, farming in the north of Makueni County is often challenging, and harsh climatic events are no exception. For that reason, several female farmers have deliberately taken several measures to increase their farms' climatic resilience. Examples are the creation of ponds to harvest rainwater from the road, adding zaï pits to the crop fields, making terraces to slow down the runoff of water, and planting more trees on the farms (Photo 3.31).

Photo 3.31 Farm in Makueni County, Kenya

Combining all these different interventions has led to significant results within a couple of years. The soil moisture levels have increased, and this had multiple benefits and resulted in better-buffered temperatures. The tree rows reduced wind speeds, which reduced evaporation and helped control wind erosion, resulting in more fertile soils. These microclimatic improvements enabled a higher and more diverse crop production. Deliberately taking these measures positively transformed these farms, and a tremendous difference can be noticed from other farms in the area that have not implemented these interventions. Here temperatures are less buffered, and adverse effects from winds are noticeable, affecting crop health.

the estimated quantified local climate impacts and the landscape characteristics required for each climate management practice of this guiding kit. There may be many more practices specific to different geographies and global regions. As we more consciously create stronger local climates, the repertoire of interventions can be expected to grow and to become more precise. A site is not limited to just one practice; multiple methods can be implemented simultaneously. Restoring and improving the local climate thus requires sensitivity to ecological nuance and the needs and aspirations of people living in the area. This underlines the importance of local knowledge from watershed managers and farmers in local climate management (McNamara and Buggy, 2017). In the end, the people who live and work in a specific environment or ecosystem know that ecosystem best and are most knowledgeable about its response to certain practices. This, together with scientific knowledge on the micrometeorological processes of the local climate system, provides robust input to selecting the best management practices. We should strive for co-design and co-innovation processes in agricultural systems.

An anecdotal example of how a combination of water conservation, regreening, and other techniques transformed the climate on farms in Makueni County, Kenya can be found in Box 3.17.

CHAPTER 4
Measuring and monitoring the local climate

It is possible to measure many of the variables that make up the micro- and mesoclimate, to see how they develop over time and behave in space. This sets the path to manage local climates at the field and landscape level.

This chapter explains various current techniques and discusses how this quantitative information can assist decision-making. The advantages and disadvantages of several methods are mentioned, and new frontiers are discussed. They should always be applied together with local knowledge, for land users themselves have a rich, experience-based understanding of local climates and the variations and changes therein.

If, together with this local citizen knowledge, we can better measure the underlying physical land surface processes, it will be feasible to relate local climates to land and water management practices. There is thus a generic need to quantify the state variables (e.g. soil moisture, air humidity, air temperature), fluxes, and processes of the soil–vegetation–atmosphere continuum and see how these respond to local climate management practices such as those covered in this book.

Seasonal climate and weather forecast from weather stations are often used by farmers to, for example, plan cropping patterns and variety, and pest and disease management. But data derived from country-wide weather-station networks does not capture local climates (Lembrechts et al., 2021a). These data sets bear little resemblance to conditions experienced in the field and miss out on the subtleties, diversity, and complexity of the local climate (Duffy et al., 2021). Local climates are highly variable due to turbulent flows caused by obstacles and surface roughness (e.g. vegetation structure and windbreaks). This means that global climate data sets (such as GLDAS, ERA5, and more) might be helpful in first-level assessments in the absence of any other information; however, their resolution is too coarse (10 km × 10 km up to 50 km × 50 km) to describe these highly local situations.

What could be the impacts of using the wrong micrometeorological information? Schultze et al. (2021) give an example of a citrus grower receiving local news that the air temperature might drop below 0°C that evening. The grower has the option to turn on sprinklers to displace the dew point within the trees and counteract any potential frost. Turning on the sprayer hoses could accidentally damage the trees if the local temperature never dropped below freezing point. This highlights the importance of farmers knowing their local conditions through data at a fine spatial resolution to act accordingly. Also, a study by Roncoli et al. (2011) conducted in Uganda showed that when farmers received seasonal climate forecasts, they were unlikely to utilize this information and adopt alternative farming strategies accordingly.

Too coarse climate information, both in spatial and temporal resolution, is thus of minimal relevance at the community level for local climate management. This shows the need to monitor and map the local climate at the farm and landscape level to support local climate management and to complement the use of coarser seasonal data. Monitoring and mapping the local climate can familiarize farmers and practitioners with this under-explored aspect of their farm or site. This will help them to recognize patterns and adapt their practices accordingly. Such knowledge-led awareness is fundamental for increasing local resilience to global climate change and for realizing effective local climate management.

As described in Chapter 2, local climates can be characterized through the following variables: solar radiation, air temperature, soil temperature, air humidity, wind direction and speed, and soil moisture. Quantifying these variables and processes, such as dew formation and precipitation, on a micro- and mesoclimate scale will advance the ability to document and study the local climatic changes resulting from specific management practices. This chapter discusses how this can be done with local stations, field sensors, and remote sensing techniques.

4.1 Local stations

Many local climate variables can be measured directly with dedicated sensors packed in a hydro-meteorological station. These stations can estimate a range of atmospheric variables, typically: air temperature, rainfall, air pressure, air humidity, wind speed, wind direction, and light intensity. This makes local stations a powerhouse that has been used to feed local data to climatic models, weather forecasts, and early warning systems. Box 4.1 provides an example of the Trans-African Hydro-Meteorological Observatory (TAHMO), a promising project to develop a dense network of hydro-meteorological monitoring stations in sub-Saharan Africa: one every 30 kilometres.

> **Box 4.1** Trans-African Hydro-Meteorological Observatory (TAHMO)
>
> The idea behind this project is to develop a dense network of hydro-meteorological monitoring stations in sub-Saharan Africa, which entails the installation of 20,000 stations across the continent. These stations can monitor various variables: short-wave radiation, wind gusts, relative humidity, atmospheric pressure, wind direction, wind speed, surface air temperature, and precipitation (https://tahmo.org).
>
> Applying innovative sensor and information and communications technology, TAHMO stations are relatively inexpensive (each station should not cost more than US$500) and robust. These local weather data will be combined with models and satellite observations to obtain insight into the distribution of water and energy stocks and fluxes (Van De Giesen, 2013). The more dense the spread of these hydro-meteorological stations over regions, the more accurate the local climate data acquired.

Hydro-meteorological stations are commonly used by hydrological agencies, irrigation districts, and farmers to measure rainfall and other climatic parameters in real-time. Unfortunately, the maintenance of public weather stations is not always optimal. Standard stations from the World Meteorological Organization usually get better maintenance and data downloads. Still, these stations only provide a point measurement that could be located 10 to 200 kilometres away from the site of interest. This means the data available is not always directly representative of the local climate of a specific location as these are highly variable. Using automated weather stations directly on the farm or local site of interest with immediate access to the data can be advantageous despite requiring investments (Photo 4.1). In areas with fewer hydro-meteorological stations, it is possible to approximate the climate variables' values. Point values may be interpolated using values from the nearest stations (Holtslag, 1984; Hutjes, 1996).

4.2 Field sensors

There have been enormous new developments in low-cost, reliable field sensors in recent years. Localized effects of farming and conservation practices can be precisely appreciated with local point measurements. In cases where changes in the field are small (e.g. an area shaded by a tree and an area in the sun), small sensors attached to data loggers are eminently suitable. These sensors can, for example, measure soil moisture and soil temperature related to various local climate improvements, for instance in areas with or without mulching, or they can be used to understand soil moisture dynamics behind small soil bunds (Box 4.2).

98 MANAGING THE LOCAL CLIMATE

Photo 4.1 Example of a weather station on a local site. Weather stations can be equipped with many sensors to conduct a broad range of measurements such as wind speed and direction, air temperature, rainfall, relative humidity, solar radiation, and temperature.

When arrays of these sensors are installed, nuances can be captured, and gradients discovered. Compared to hydro-meteorological stations, a benefit of these sensors is that they can also measure soil variables. Moreover, an increasing number of studies are demonstrating that even some of the cheaper soil moisture sensors, after careful data manipulation, can provide reliable data (García et al., 2020; Trilles, 2020; Placidi et al., 2021). In recent years there has been a lot of development in the sector, and the prospect for more precise and cheap sensor setups is promising (García et al., 2020). An example is the newly developed local climate data logger called the temperature-moisture sensor that integrates sensors for measuring air, surface and soil temperature, and soil moisture into a compact unit with a long-lasting battery and large memory capacity (Wild et al., 2019).

> **Box 4.2** Field sensors to detect the microclimatic effects of soil bunds
>
> As part of the Green Future Farming project carried out by MetaMeta, Justdiggit, and RAIN, several soil and water conservation structures were implemented in a plot called the Meshenani Grass Seed Bank in the Amboseli National Park, Kenya. The plot is also seeded with native grass seeds for regreening purposes. These interventions aim to increase soil moisture and vegetation levels on the field.
>
> To monitor changes in the local climate resulting from these interventions, 10 temperature-moisture sensor loggers have been placed throughout the plot (Photo 4.2). These sensors can collect soil temperature data at a depth of 8 centimetres, the soil surface temperature, and the air temperature at the height of 45 centimetres. This allows for precise monitoring of the microclimatic changes.
>
>
>
> **Photo 4.2** The temperature-moisture sensor placed in the Meshenani Grass Seed Bank
> *Credit*: Tijmen Schults

Small field sensors can help farmers make sense of their environment and act and tweak practices accordingly to influence the local climate positively. For instance, a series of moisture sensors can be applied to check the effectiveness of various mulching materials and water conservation measures to retain soil moisture. Temperature sensors can check how air temperature changes under different shading gradients. The use of soil moisture sensors to steer irrigation supplies is already in place by small farmers in an increasing number of sites (see Box 4.3).

Box 4.3 Remote control farming in Cambodia

Im Naisreang is a 60-year-old commercial farmer in the Preah Vihear region in Cambodia. She lives with her husband and grows different crops on their 4.5 hectare farm. Im adapted a soil moisture alarm system to her farm, which helps her irrigate crops on time. The soil moisture sensor was connected to the solar pump panel, which provides information related to soil moisture percentage. Based on this reliable information, she can decide when to irrigate her crops, thus saving lots of water. This proved to be an efficient irrigation system that she controls remotely, and it helps her save both water and time (Lim and Mukherjee, 2021).

Photo 4.3 Im Naisreang with her phone, which she uses to irrigate her farm
Credit: Lim and Mukherjee, 2021

Lembrechts and Lenoir (2020) have taken up the task to initiate a shared repository of microclimatic data retrieved from sensors from all over the world into a global geo-database. This database initiative is called SoilTemp and will allow patterns in soil temperature and the boundary layer to be related to what is happening in the upper layers and enable calibration and validation of global mechanistic models of soil temperature and (micro) climate. Lembrechts et al. (2021b) generated global gridded maps of below-canopy and near-surface soil temperature at 1-kilometre resolution from this database.

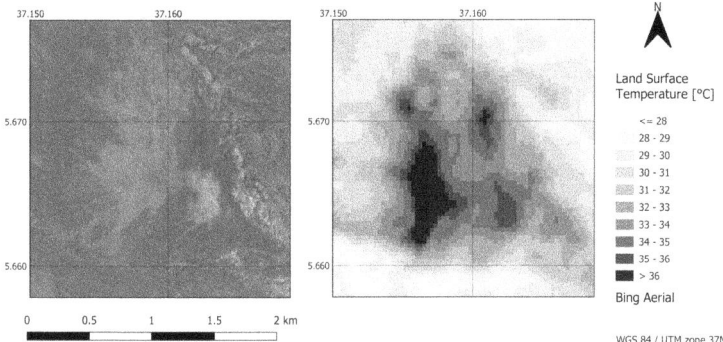

Figure 4.1 Spatial variation in land surface temperature (LST) in Bole Dugulo, Ethiopia on 10 January 2022. Here it can be seen that a lower tree density (left map) corresponds to a higher LST (right map). The LST is derived from Landsat 8 satellite imagery using the Google Earth Engine script by Ermida et al. (2020)
Source: Map created by Tijmen Schults

4.3 Remote sensing

Remote sensing is the science of obtaining and interpreting information about an area's physical properties, state variables, and water/energy fluxes from a distance based on spectral radiative measurements. Remote sensing data can be obtained through satellites, aircraft, and drones. The advantage of satellite remote sensing is that large areas can be easily covered, and variations can be easily detected. Figure 4.1 shows an example.

In particular, stationary satellite systems have a fine temporal resolution and measure solar radiation and surface temperature variations over the day. As remote sensing data is archived, we can also contract time series, detect trends and changes, and see the impact of improvements or land degradations. There are also systems where such historical data have been collected and pre-processed, facilitating time analysis.

Geostationary satellites such as Meteosat Second Generation (MSG) and Geostationary Operational Environmental Satellite (GOES) measure solar radiation at a resolution of 2 to 3 kilometres. A second well-known contribution from satellites is measuring land surface temperature (LST) (e.g. Sobrino et al., 2020), which geostationary satellites measure with small time steps.

Sun-synchronous satellites also measure LST. The Sentinel-3 satellites and MODIS do this at 1,000-metre resolution at a time interval of 1 day. At time intervals of 8 days, Landsat-8 measures LST at 100-metre spatial resolution.

In addition, the International Space Station (ISS) is equipped with a thermal sensor referred to as ECOSTRESS (Silvestri et al., 2020). ECOSTRESS is mapping the LST with a spatial detail of 38 to 69 metres. Due to the orbit of the ISS, the time of acquisition changes between consecutive overpasses. These operating systems allow measurement of the land surface conditions locally and are thus highly suitable for quantifying local climates.

The research community has taken up the availability of daily LST maps at various spatial resolutions to bring this data to the next level: the surface energy balance. Various energy balance algorithms have been developed in the last 30 years, ranging from simple to advanced (Courault et al. 2005; Kalma et al. 2008). Energy balances describe how much solar energy and long-wave energy are converted into latent heat flux (see Chapter 2). This latent heat flux can be re-expressed into the actual evapotranspiration rate using physical constants. The advantage of such a technique is that all complex hydrological processes that determine evapotranspiration can be circumvented: evapotranspiration results from certain energy flows instead of complex water distribution processes and transient soil moisture flows.

A main model tested and validated in more than 45 countries by local universities and agricultural research stations is the Surface Energy Balance Algorithm for Land (SEBAL) (Box 4.4). The complete SEBAL algorithm was published in 1998 (Bastiaanssen et al., 1998). SEBAL can be used by the international community free of charge. It can also be implemented in the Google Earth Engine environment (Laipelt et al., 2021). OpenET is a new initiative from California that provides an option to run seven different energy balance models simultaneously and comes up with an ensemble prediction of evapotranspiration with

---

Box 4.4 Surface Energy Balance Algorithm for Land (SEBAL)

The Surface Energy Balance Algorithm for Land (SEBAL) estimates aspects of the hydrological cycle by using the surface energy balance. By using satellite data as an input, SEBAL quantifies the energy balance. It first needs to determine the net radiation (R_n). The net radiation is the difference between the incoming and outgoing radiation fluxes. SEBAL is based on the principle that this net radiation contributes to either evaporation processes (latent heat flux) or heating the soil and air (sensible heat flux). If this sensible heat flux is subtracted from the net radiation, the residual is the evapotranspiration. Thus, using the energy balance and satellite data as an input, SEBAL maps evapotranspiration, biomass growth, soil moisture, and soil temperature. This can also be done at a field scale that is most relevant for the agricultural sector as well as the landscape level that describes the mesoclimate.

a spatial detail of 30 metres (www.openetdata.org). SEBAL is one of these seven models. This is a significant step forward to getting data on crop evapotranspiration within the field. OpenET is, at the time of writing, only available for western US states.

Crop and landscape evapotranspiration can be a dominant factor in the water balance of irrigated fields and catchments. This large water consumption is sometimes considered undesirable because it increases the footprint of several modern goods (Hoekstra and Mekonnen, 2012). However, the ecosystem services from crops, wetlands, and green landscapes should be considered simultaneously. While this book has reported on windbreaks, soil conservation, reduced surface runoff, and more, it should be recognized that the earth system would have been much hotter without vegetation and evaporative cooling. We, therefore, propose that water footprinting should also be evaluated from the viewpoint of ecosystem services and not be limited to consumables production. If the energy contributing to latent heat were instead sensible heat in the absence of vegetation, air at observation height would have been much warmer. The midday air temperature cooling due to evapotranspiration and shade can be computed from the equation for sensible heat transfer. This regulating impact of green landscapes on local climate and, in general, on global warming is often overlooked by the lack of a proper measurement system. Land surface measurements on daily recurring satellites and the processing with surface energy balance models is a significant step forward for mapping the local climate. Furthermore, knowledge of spatially distributed latent heat flux at a 10-metre resolution can be used in an analogue manner to describe very local atmospheric vapour pressure and absolute air humidity.

There have been significant breakthroughs in quantitatively describing local climates using thermal remote sensing in recent years. For example, Maclean et al. (2019) developed a modelling package for the meso- and microclimate to determine the range in near-ground air temperatures using an energy balance equation. In follow-up research, Maclean (2019) presents methods for providing fine-grained, hourly and daily estimates of the current and future temperature and soil moisture over decadal timescales. Huang et al. (2020) estimated soil temperature from LST data on a 1-kilometre scale. Hooker et al. (2018) developed a statistical model to predict air temperature from remotely sensed LST. They used it to estimate a novel data set of monthly air temperature at the spatial resolution of 5.55 kilometres. Rawat et al. (2019) developed a methodology to retrieve soil moisture data from an agricultural area using the Landsat-8, Sentinel-1 satellite data and Modified Water Cloud Model.

Remote sensing satellite data can also compute the normalized difference vegetation index and the normalized difference water index. The first is used to estimate the vegetation density and health. The latter can be used to monitor changes in the water content of leaves or water bodies. Wang et al. (2020) compared several satellites and discovered that Landsat-8 was best suited for the water index and Sentinel-2 best suited for the vegetation index.

The number of satellite systems increases every year with a continuous improvement of their spatial and temporal resolution – moreover, the number of available bands in these satellites increases, resulting in new and valuable information. The reliability of this remote sensing satellite data also increases due to computational and technological improvements. Scientific advances to downscale the LST data from satellites have also been made (Yang et al., 2019; Zawadzka et al., 2020).

An important new satellite frontier is the launching of low-flying satellites. Planet Labs ('Planet') has a large fleet of active nanosatellites in orbit at a height of about 450 kilometres, offering an unprecedented monitoring capacity of daily and global RGB (red, green, and blue) and NIR (near infrared) image capture at 3–5-metre resolution.

In addition to satellites, lightweight drones (Photo 4.4) are particularly suited to measure local climate factors because of high-resolution imagery. This high resolution makes drones and platforms the best fit for mapping small fields. Detailed vegetation

Photo 4.4 Example of a drone flying over an agricultural field

structure data in terms of vegetation height and width is crucial to better account for the effects of wind on local climate. Canopy surface roughness and vertical structures on a site improve wind modelling and offer promising opportunities to make more accurate predictions of the near-surface wind fields (De Vries et al., 2003). Airborne LiDAR (light detection and ranging)-derived maps are also suitable to analyse the fine-scale variation of soil moisture, air temperature (Kašpar et al., 2021), and air humidity (Zellweger et al., 2019). Thermal cameras on drones are very suitable to map the local climate within fields up to the spatial level of a single leaf. The number of applications with thermal cameras on drones is increasing (e.g. Sagan et al., 2019). Increased efforts to use remote sensing to upscale in situ local climate measurements will further our mechanistic understanding of how topography and vegetation structure determine the local climate (Zellweger et al., 2019). The detailed spatial and temporal fine resolution data derived from drone remote sensing may lead to more realistic predictions of the local climate, assisting local climate management.

Combining data collected from both remote sensing methods and field measurements is a most productive way to assess and understand the local climate at relevant spatio-temporal scales (Liaghat and Balasundram, 2010). Integrating in situ measurements with remote sensing data from multiple platforms infers fine-scale temperature differences over larger spatial extents than achievable solely with field measurements or drone data (Reichenau et al., 2016). The different sources complement one another. Kašpar et al. (2021) provide an example of how unmanned aerial systems can substitute ground canopy measurements for air temperature modelling. Another example is provided by Beck et al. (2019) that merges gauges, satellite, and reanalysis data to provide precipitation data hourly on an 11.1-kilometre scale. In addition, it is strongly recommended to combine these measurements with the experiential knowledge of local land users.

Appendix A provides an overview of measurement tools discussed in this chapter with their advantages, disadvantages, and cost factors.

Developments in local climate modelling using the mentioned tools (local stations, field sensors, remote sensing, and thermal drone data) provide breakthroughs for improved local climate management. A paradigm shift is happening towards more fine-grained estimates of local climate variables, and a myriad of new techniques and technologies are now available, ranging from in situ logging to thermal remote sensing. This opens the door for a bright local climatic future, especially when mechanistic models are combined with in situ microclimatic measurements (Lembrechts and Lenoir, 2020). Increased precision in local climate modelling and mapping allows for profound knowledge of the workings of the local climate system – at both farm and landscape levels. There is a world to be discovered here, and more profound knowledge can assist better decision-making in local climate management.

CHAPTER 5
Conclusion: A patchwork of local action

> *We stop anyone from chopping trees and make them understand that it will disturb the environmental balance*
> Meena Devi, farmer in Jharkhand, India

We can help create more conducive local climates intentionally by proactive, well-considered, and measurable action. This results in local climates that are cooler and better buffered against extreme events. We can do this area by area, creating a mosaic of stronger local climates, all combined contributing to 'making a new climate', as Judith Schwartz (2020) proposed in *The Reindeer Chronicles*. Managing local climates is a complementary way to address global climate change, next to mitigation and adaptation. It is based on well-considered local action – in retaining water, regreening, and landscape planning. It can be done farm by farm and landscape by landscape.

Several of the measures that can be undertaken to regulate local climates at the micro-and meso-level are described in this book. Still, many more possible measures are appropriate to different geographies and temperature zones. In many cases, these measures not only serve to create conducive climate conditions, but they are productive assets in their own right or serve other purposes. This makes it possible to improve local climate with ingenuity and clever designs of water bodies, land management measures, and regreening rather than by massive investments necessarily. There is much to research and understand yet on the different elements that make up the local climate, such as the interaction and effects of various measures on local temperature, wind, humidity, and rainfall.

By combining remote sensing, new generation field sensors, and local weather stations, we have a powerful and still evolving instrument to monitor local climate over time and in space – at both plot and landscape level – and plan the measures. Where the systematic landscape-wide implementation of these measures has been achieved, the cooling effect has been from 1.5 to 2.5°C, sometimes even more. This is in the bandwidth of predicted global climate change.

We may also think of the world's climate as a patchwork of local climates and our human agency as a patchwork of local actions (Photo 5.1), carefully stitched together and combined to create fields,

108 MANAGING THE LOCAL CLIMATE

Photo 5.1 Aerial view of the bocage landscape in La Vendelée, Manche, in France. The bocage landscape is characterized by woodland structures, dividing the pastoral land into smaller plots with separated, improved local climates. This photo displays how improving the local climate can be done area by area, consequently covering an entire landscape as a patchwork

Photo 5.2 Maasai women digging a soil bund for water conservation in the Meshenani Grass Seed Bank as part of the Green Future Farming project

landscapes, regions, and continents that are resilient to climate change. Not only that, but to have better conditions for farming, livestock keeping, pest and disease control, biodiversity, drought resilience, public health, and human endeavour.

Local climate management can be a powerful frontier in smoothening out the impact of global climate change and creating a mosaic of stronger agroecological systems.

References

Acharya, C.L., Hati, K.M., Bandyopadhyay, K.K. and Daniel, H. (2005) 'Mulches', in D. Hillel (ed.), *Encyclopedia of Soils in the Environment*, pp. 521–32, Academic Press, Cambridge.

Adeboye, O.B., Schultz, B., Adekalu, K.O. and Prasad, K. (2017) 'Soil water storage, yield, water productivity and transpiration efficiency of soybeans (*Glyxine max* L. Merr) as affected by soil surface management in Ile-Ife, Nigeria', *International Soil and Water Conservation Research* 5(2): 141–50 <https://doi.org/10.1016/j.iswcr.2017.04.006>.

Adekalu, K.O., Balogun, J.A., Aluko, O.B., Okunade, D.A., Gowing, J.W. and Faborode, M.O. (2009) 'Runoff water harvesting for dry spell mitigation for cowpea in the savannah belt of Nigeria', *Agricultural Water Management* 96(11): 1502–8 <https://doi.org/10.1016/j.agwat.2009.06.005>.

Adimassu, Z., Alemu, G. and Tamene, L. (2019) 'Effects of tillage and crop residue management on runoff, soil loss and crop yield in the humid highlands of Ethiopia', *Agricultural Systems* 168: 11–18 <https://doi.org/10.1016/j.agsy.2018.10.007>.

Aerts, R., Wagendorp T., November E., Behailu M., Deckers J. and Muys B. (2004) 'Ecosystem thermal buffer capacity as an indicator of the restoration status of protected areas in the northern Ethiopian highlands', *Restoration Ecology* 12: 586–96 <https://doi.org/10.1111/j.1061-2971.2004.00324.x>.

Ahmed, M. and Al-Dousari, A. (2014) 'Application of water harvesting system for rehabilitation of degraded lands in Liyah in northern Kuwait', *Journal of Arid Land Studies* 24: 257–60.

Ahrends, A., Hollingsworth, P.M., Beckschäfer, P., Chen, H., Zomer, R.J., Zhang, L., Wang, M. and Xu, J. (2017) 'China's fight to halt tree cover loss', *Proceedings of the Royal Society B: Biological Sciences* 284 (1854): 20162559 <https://doi.org/10.1098/rspb.2016.2559>.

Allaby, M. (2015) *The Gardener's Guide to Weather and Climate: How to Understand the Weather and Make It Work for You*, Timber Press, Portland, OR.

Allen, R.G., Pereira, L.S., Raes, D. and Smith, M. (1998) *Crop Evapotranspiration: Guidelines for Computing Crop Water Requirements*, FAO Irrigation and Drainage Paper 56, FAO, Rome.

Alter, R.E., Im, E.S. and Eltahir, E.A. (2015) 'Rainfall consistently enhanced around the Gezira Scheme in East Africa due to irrigation', *Nature Geoscience* 8(10): 763–7 <https://doi.org/10.1038/ngeo2514>.
Altieri, M.A. (1996) 'Indigenous knowledge re-valued in Andean agriculture', *ILEIA Newsletter* 12(1): 7.
Amare, T., Zegeye, A.D., Yitaferu, B., Steenhuis, T.S., Hurni, H. and Zeleke, G. (2014) 'Combined effect of soil bund with biological soil and water conservation measures in the northwestern Ethiopian highlands', *Ecohydrology & Hydrobiology* 14(3): 192–9 <https://doi.org/10.1016/j.ecohyd.2014.07.002>.
Angelo, D., Morales, E. and Programa de Suka Kollus (2008) *Suka Kollus; una tecnología ancestral para el tiempo actual*. La Paz: Prosuko.
Antle, J.M., Basso, B., Conant, R.T., Godfray, H.C.J., Jones, J.W., Herrero, M., Howitt, R.E., Keating, B.A., Munoz-Carpena, R., Rosenzweig, C., Tittonell, P. and Wheeler, T.R. (2017) 'Towards a new generation of agricultural system data, models and knowledge products: design and improvement', *Agricultural Systems* 155: 255–68 <https://doi.org/10.1016/j.agsy.2016.10.002>.
Araya, A. and Stroosnijder, L. (2010) 'Effects of tied ridges and mulch on barley (*Hordeum vulgare*) rainwater use efficiency and production in northern Ethiopia', *Agricultural Water Management* 97(6): 841–7 <https://doi.org/10.1016/j.agwat.2010.01.012>.
Aubertin, G.M. (1971) *Nature and Extent of Macropores in Forest Soils and their Influence on Subsurface Water Movement*, Vol. 192, Res. Pap. NE-192, Northeastern Forest Experiment Station, Upper Darby, PA.
Aubertin, G.M. and Peters, D.B. (1961) 'Net radiation determinations in a cornfield 1', *Agronomy Journal* 53(4): 269–72 <https://doi.org/10.2134/agronj1961.00021962005300040019x>.
Barry, R. and Blanken, P. (2016) 'Microclimatic elements', in *Microclimate and Local Climate*, pp 11–51, Cambridge University Press, Cambridge <https://doi.org/10.1017/CBO9781316535981.004>.
Bastiaanssen, W.G., Menenti, M., Feddes, R.A. and Holtslag, A.A.M. (1998) 'A remote sensing surface energy balance algorithm for land (SEBAL). 1. Formulation', *Journal of Hydrology* 212: 198–212 <https://doi.org/10.1016/S0022-1694(98)00253-4>.
Beck, H., Wood, E., Pan, M., Fisher, C.K., Miralles, D.G., van Dijk, A.I.J.M., McVicar, T.R. and Adler, R.F. (2019) 'MSWEP V2 global 3-hourly 0.1 precipitation: methodology and quantitative assessment', *Bulletin of the American Meteorological Society* 100(3): 473–500 <https://doi.org/10.1175/BAMS-D-17-0138.1>.
Benítez, Á., Prieto, M. and Aragón, G. (2015) 'Large trees and dense canopies: key factors for maintaining high epiphytic diversity on trunk bases (bryophytes and lichens) in tropical montane forests',

Forestry: An International Journal of Forest Research 88(5): 521–7 <https://doi.org/10.1093/forestry/cpv022>.

Blevins, R.L., Cook, D., Phillips, S.H. and Phillips, R.E. (1971) 'Influence of no tillage on soil moisture 1', *Agronomy Journal* 63(4): 593–6 <https://doi.org/10.2134/agronj1971.00021962006300040024x>.

Boerma, D. (2013) Waru Waru (or raised-bed) agriculture is a technology developed over centuries in the Peruvian Andes. Agricultures Network. <http://www.agriculturesnetwork.org/resources/learning/mod2-online/edu-res/r3/waru-waru-peru/view> [accessed 5th May 2022; website no longer active].

Bonan, G.B. (2016) *Ecological Climatology*, 3rd edn, Cambridge University Press, Cambridge.

Borger, C.P.D., Hashem, A. and Powles, S.B. (2016) 'Manipulating crop row orientation and crop density to suppress *Lolium rigidum*', *Weed Research* 56: 22–30 <https://doi.org/10.1111/wre.12180>.

Borgia, C. (2017) '"Moon water harvesting" for volcanic wines in Lanzarote', TheWaterChannel, 3 October [blog] <https://thewaterchannel.tv/thewaterblog/moon-water-harvesting-for-volcanic-wines-in-lanzarote/> [accessed 10 November 2021].

Bouchet, R.J. (1963) 'Evapotranspiration réelle evapotranspiration potentielle, signification climatique', in *International Association of Scientific Hydrology, General Assembly of Berkeley, Transactions*, Vol. 2, *Evaporation*, pp. 134–42, Berkeley, CA.

Bowen, I.S. (1926) 'The ratio of heat losses by conduction and by evaporation from any water surface', *Physical Review* 27(6): 779.

Bristow, K.L. (1988) 'The role of mulch and its architecture in modifying soil temperature', *Australian Journal of Soil Research* 26: 269–80 <https://doi.org/10.1071/SR9880269>.

Brown, K.W. and Rosenberg, N.J. (1972) 'Shelter effects on microclimate, growth and water use by irrigated sugar beets in the Great Plains', *Agriculture for Meteorology* 9: 241–63 <https://doi.org/10.1016/0002-1571(71)90025-2>.

Brown, R.D. (2011) 'Ameliorating the effects of climate change: modifying microclimates through design', *Landscape and Urban Planning* 100(4): 372–4 <https://doi.org/10.1016/j.landurbplan.2011.01.010>.

Brutsaert, W. (1982) *Evaporation into the Atmosphere: Theory, History and Applications*, Springer Science & Business Media, Berlin/Heidelberg.

Buchele, W.F., Collins, E.V. and Lovely, W.G. (1955) 'Ridge farming for soil and water control', *Agricultural Engineering* 36: 324–31 <https://dr.lib.iastate.edu/handle/20.500.12876/1776>.

Burrows, W.C. (1963) 'Characterization of soil temperature distribution from various tillage-induced microreliefs', *Soil Science Society of*

America Journal 27(3): 350–3 <https://doi.org/10.2136/sssaj1963.03615995002700030038x>.

Busari, M.A., Kukal, S.S., Kaur, A., Bhatt, R. and Dulazi, A.A. (2015) 'Conservation tillage impacts on soil, crop and the environment', *International Soil and Water Conservation Research* 3(2): 119–29 <https://doi.org/10.1016/j.iswcr.2015.05.002>.

Caborn, J.M. (1957) *Shelterbelts and Microclimate (No. 29)*, HM Stationery Office, London.

Campi, P., Palumbo, A.D. and Mastrorilli, M. (2009) 'Effects of tree windbreak on microclimate and wheat productivity in a Mediterranean environment', *European Journal of Agronomy* 30(3): 220–7 <https://doi.org/10.1016/j.eja.2008.10.004>.

Carter, M.R. (2005) 'Conservation tillage', in D. Hillel (ed.), *Encyclopedia of Soils in the Environment*, pp. 521–32, Academic Press, Cambridge.

Castelli, G., Castelli, F. and Bresci, E. (2018) 'Evidence of climate mitigation from landscape restoration and water harvesting: a remote sensing approach', *AGU Fall Meeting Abstracts* 2018: GC33F-1427.

Castelli, G., Castelli, F. and Bresci, E. (2019a) 'Mesoclimate regulation induced by landscape restoration and water harvesting in agroecosystems of the Horn of Africa', *Agriculture, Ecosystems & Environment* 275: 54–64 <https://doi.org/10.1016/j.agee.2019.02.002>.

Castelli, G., Oliveira, L.A.A., Abdelli, F., Dhaou, H., Bresci, E. and Ouessar, M. (2019b) 'Effect of traditional check dams (jessour) on soil and olive trees water status in Tunisia', *Science of The Total Environment* 690: 226–36 <https://doi.org/10.1016/j.scitotenv.2019.06.514>.

Chai, Q., Gan, Y., Turner, N.C., Zhang, R.Z., Yang, C., Niu, Y. and Siddique, K.H. (2014) 'Water-saving innovations in Chinese agriculture', *Advances in Agronomy* 126: 149–201 <https://doi.org/10.1016/B978-0-12-800132-5.00002-X>.

Chen, Y., Liu, S., Li, H., Li, X.F., Song, C.Y., Cruse, R.M. and Zhang, X.Y. (2011) 'Effects of conservation tillage on corn and soybean yield in the humid continental climate region of Northeast China', *Soil and Tillage Research* 115: 56–61 <https://doi.org/10.1016/j.still.2011.06.007>.

Cleugh, H.A. (1998) 'Effects of windbreaks on airflow, microclimates, and crop yields', *Agroforestry Systems* 41(1): 55–84 <https://doi.org/10.1023/A:1006019805109>.

Correggiari, M., Castelli, G., Bresci, E. and Salbitano, F. (2017) 'Fog collection and participatory approach for water management and local development: practical reflections from case studies in the Atacama Drylands', in M. Ouessar, D. Gabriels, A. Tsunekawa, and S. Evett (eds), *Water and Land Security in Drylands*, Springer, Cham <https://doi.org/10.1007/978-3-319-54021-4_14>.

Courault, D., Seguin, B. and Olioso, A. (2005) 'Review on estimation of evapotranspiration from remote sensing data: from empirical to numerical modeling approaches', *Irrigation and Drainage Systems* 19(3-4): 223–49 <https://doi.org/10.1007/s10795-005-5186-0>.

Covell, S., Ellis, R.H., Roberts, E.H. and Summerfield, R.J. (1986) 'The influence of temperature on seed germination rate in grain legumes: I. A comparison of chickpea, lentil, soyabean and cowpea at constant temperatures', *Journal of Experimental Botany* 37(5): 705–15 <https://doi.org/10.1093/jxb/37.5.705>.

Critchley, W., Siegert, K., Chapman, C. and Finket, M. (2013) *Water Harvesting: A Manual for the Design and Construction of Water Harvesting Schemes for Plant Production*, Scientific Publishers, Rajasthan.

Crompton, O., Corre, D., Duncan, J.. and Thompson, S. (2021) 'Deforestation-induced surface warming is influenced by the fragmentation and spatial extent of forest loss in Maritime Southeast Asia', *Environmental Research Letters* 16(11): 114018 <https://doi.org/10.1088/1748-9326/ac2fdc>.

Dai, A. (2013) 'Increasing drought under global warming in observations and models', *Nature Climate Change* 3(1): 52–8 <https://doi.org/10.1038/nclimate1633>.

Dass, A., Sudhishri, S., Lenka, N.K. and Patnaik, U.S. (2011) 'Runoff capture through vegetative barriers and planting methodologies to reduce erosion, and improve soil moisture, fertility and crop productivity in southern Orissa, India', *Nutrient Cycling in Agroecosystems* 89(1): 45–57 <https://doi.org/10.1007/s10705-010-9375-3>.

Davis, K.T., Dobrowski, S.Z., Holden, Z.A., Higuera, P.E. and Abatzoglou, J.T. (2019) 'Microclimatic buffering in forests of the future: the role of local water balance', *Ecography* 42(1): 1–11 <https://doi.org/10.1111/ecog.03836>.

De Decker, K. (2020) 'Fruit trenches: cultivating subtropical plants in freezing temperatures', *Low-Tech Magazine* [online] <https://www.lowtechmagazine.com/2020/04/fruit-trenches-cultivating-subtropical-plants-in-freezing-temperatures.html> [accessed 21 July 2022].

De Frenne, P., Rodríguez-Sánchez, F., Coomes, D.A., Baeten, L., Verstraeten, G., Vellend, M., Bernhardt-Römermann, M., Brown, C.D., Brunet, J., Cornelis, J., Decocq, G.M., Dierschke, H., Eriksson, O., Gilliam, F.S., Hédl, R., Heinken, T., Hermy, M., Hommel, P., Jenkins, M.A., Kelly, D.L., Kirby, K.J., Mitchell, F.J.G., Naaf, T., Newman, M., Peterken, G., Petřík, P., Schultz, J., Sonnier, G., Van Calster, H., Waller, D.M., Walther, G-R., White, P.S., Woods, K.D., Wulf, M. Jessen Graae, B. and Verheyen, K. (2013) 'Microclimate moderates plant responses to macroclimate warming', *Proceedings*

of the National Academy of Sciences 110(46): 18561–5 <https://doi.org/10.1073/pnas.1311190110>.
De Frenne, P., Lenoir, J., Luoto, M., Scheffers, B.R., Zellweger, F., Aalto, J., Ashcroft, M.B., Christiansen, D.M., Decocq, G., De Pauw, K., Govaert, S., Greiser, C., Gril, E., Hampe, A., Jucker, T., Klinges, D.H., Koelemeijer, I.A., Lembrechts, J.J., Marrec, R., Meeussen, C., Ogée, J., Tyystjärvi, V., Vangansbeke, P. and Hylander, K. (2021) 'Forest microclimates and climate change: importance, drivers and future research agenda', *Global Change Biology* 27(11): 2279–97 <https://doi.org/10.1111/gcb.15569>.
De Vries, A.C., Kustas, W.P., Ritchie, J.C., Klaassen, W., Menenti, M., Rango, A. and Prueger, J.H. (2003) 'Effective aerodynamic roughness estimated from airborne laser altimeter measurements of surface features', *International Journal of Remote Sensing* 24(7): 1545–58 <https://doi.org/10.1080/01431160110115997>.
Dhaliwal, L.K., Buttar, G.S., Kingra, P.K., Singh, S. and Kaur, S. (2019) 'Effect of mulching, row direction and spacing on microclimate and wheat yield at Ludhiana', *Journal of Agrometeorology* 21(1): 42–5 <https://doi.org/10.54386/jam.v21i1.202>.
Duffy, J.P. anderson, K., Fawcett, D., Curtis, R.J. and Maclean, I.M. (2021) 'Drones provide spatial and volumetric data to deliver new insights into microclimate modelling', *Landscape Ecology* 36: 685–702 <https://doi.org/10.1007/s10980-020-01180-9>.
Earle, S. (2019) *Physical Geology*, 2nd edn, BCcampus Open Education.
Ekka, S.A., Rujner, H., Leonhardt, G., Blecken, G.T., Viklander, M. and Hunt, W.F. (2021) 'Next generation swale design for stormwater runoff treatment: a comprehensive approach', *Journal of Environmental Management* 279: 111756 <https://doi.org/10.1016/j.jenvman.2020.111756>.
El-Halim, A.E.H.A. and El-Razek, U.A.E.H. (2014) 'Effect of different irrigation intervals on water saving water productivity and grain yield of maize (*Zea mays* L.) under the double ridge-furrow planting technique', *Archives of Agronomy and Soil Science* 60(5): 587–96 <https://doi.org/10.1080/03650340.2013.825900>.
Ellison, D., Futter, M.N. and Bishop, K. (2012) 'On the forest cover-water yield debate: from demand- to supply-side thinking', *Global Change Biology* 18(3): 806–20 <https://doi.org/10.1111/j.1365-2486.2011.02589.x>.
Ellison, D., Morris, C.E., Locatelli, B., Sheil, D., Cohen, J., Murdiyarso, D., Gutierrez, V., van Noordwijk, M., Creed, I.F., Pokorny, J., Gaveau, D., Spracklen, D.V., Bargués Tobella, A., Ilstedt, U., Teuling, A.J., Gebreyohannis Gebrehiwot, S., Sands, D.C., Muys, B. and Sullivan, C.A. (2017) 'Trees, forests and water: cool insights for a hot world',

Global Environmental Change 43: 51–61 <http://dx.doi.org/10.1016/j.gloenvcha.2017.01.002>.

Ermida, S.L., Soares, P., Mantas, V., Göttsche, F.M. and Trigo, I.F. (2020) 'Google Earth Engine open-source code for land surface temperature estimation from the Landsat series', *Remote Sensing* 12(9): 1471 <https://doi.org/10.3390/rs12091471>.

Evans, T.P. and Winterhalder, B. (2000) 'Modified solar insolation as an agronomic factor in terraced environments', *Land Degradation & Development* 11(3): 273–87 <https://doi.org/10.1002/1099-145X(200005/06)11:3<273::AID-LDR384>3.0.CO;2-J>.

Ewers, R.M. and Banks-Leite, C. (2013) 'Fragmentation impairs the microclimate buffering effect of tropical forests', *PLOS one* 8(3): e58093 <https://doi.org/10.1371/journal.pone.0058093>.

Farahani, S.S., Soheili, F.F. and Asoodar, M.A. (2016) 'Effects of contour farming on runoff and soil erosion reduction: a review study', *Elixir Agriculture* 101: 44089–93.

Feng, W., Yang, F., Cen, R., Liu, J., Qu, Z., Miao, Q. and Chen, H. (2021) 'Effects of straw biochar application on soil temperature, available nitrogen and growth of corn', *Journal of Environmental Management* 277: 111331 <https://doi.org/10.1073/pnas.1311190110>.

Foken, T. (2008) *Micrometeorology*, Springer-Verlag, Heidelberg.

Food and Agriculture Organization of the United Nations (FAO) (2018) *The 10 Elements of Agroecology: Guiding the Transition to Sustainable Food and Agricultural Systems* [online], Rome: FAO <http://www.fao.org/3/i9037en/i9037en.pdf> [accessed 21 July 2022].

Forman, R.T. and Baudry, J. (1984) 'Hedgerows and hedgerow networks in landscape ecology', *Environmental Management* 8(6): 495–510 <https://doi.org/10.1007/BF01871575>.

Francis, R. and Weston, P. (2015) *The Social, Environmental and Economic Benefits of Farmer Managed Natural Regeneration (FMNR)*, World Vision Australia, Melbourne, pp. 6–23.

Franks, P.J., Cowan, I.R. and Farquhar, G.D. (1997) 'The apparent feedforward response of stomata to air vapour pressure deficit: information revealed by different experimental procedures with two rainforest trees', *Plant, Cell & Environment* 20(1): 142–5 <https://doi.org/10.1046/j.1365-3040.1997.d01-14.x>.

Friedrich, K., Mölders, N. and Tetzlaff, G. (2000) 'On the influence of surface heterogeneity on the Bowen-ratio: a theoretical case study', *Theoretical and Applied Climatology* 65(3): 181–96 <https://doi.org/10.1007/s007040070043>.

Fritschen, L.J. and Van Bavel, C.H.M. (1962) 'Energy balance components of evaporating surfaces in arid lands', *Journal of*

Geophysical Research 67(13): 5179–85 <https://doi.org/10.1029/JZ067i013p05179>.

Funk, R. and Engel, W. (2015) 'Investigations with a field wind tunnel to estimate the wind erosion risk of row crops', *Soil and Tillage Research* 145: 224–32 <https://doi.org/10.1016/j.still.2014.09.005>.

García, L., Parra, L., Jimenez, J.M., Lloret, J. and Lorenz, P. (2020) 'IoT-based smart irrigation systems: an overview on the recent trends on sensors and IoT systems for irrigation in precision agriculture', *Sensors* 20(4): 1042 <https://doi.org/10.3390/s20041042>.

Gardner, C.M., Laryea, K.B. and Unger, P.W. (1999) *Soil Physical Constraints to Plant Growth and Crop Production*, Land and Water Development Division, Food and Agriculture Organization, Rome.

Gebreegziabher, T., Nyssen, J., Govaerts, B., Getnet, F., Behailu, M., Haile, M. and Deckers, J. (2009) 'Contour furrows for in situ soil and water conservation, Tigray, Northern Ethiopia', *Soil and Tillage Research* 103(2): 257–64 <https://doi.org/10.1016/j.still.2008.05.021>.

Gedamu, M.T. (2020) 'Soil degradation and its management options in Ethiopia: a review', *International Journal of Research and Innovations in Earth Science* 7(5): 59–76.

Gegner, S., Overstreet, L., DeSutter, T., Casey, F. and Cattanach, N. (2008) 'Effects of row orientation and tillage on sugar beet yield and quality II. Soil temperature, moisture, and seedling emergence', in *2008 Sugar Beet Research and Extension Reports* 39 [online] <https://www.sbreb.org/wp-content/uploads/2018/09/SeedlingEmergence.pdf> [accessed 2 August 2022].

Geiger, R., Aron, R.H. and Todhunter, P. (2003) *The Climate Near the Ground*, Rowman & Littlefield, Lanham, MD.

Gliessman, S.R. (2015) *Agroecology: The Ecology of Sustainable Food Systems*, 3rd edn, Taylor & Francis Group, Abingdon, UK.

Grace, J. (1977) *Plant Response to Wind*, Vol. 13, Academic Press, Cambridge.

Graf, A., Kuttler, W. and Werner, J. (2008) 'Mulching as a means of exploiting dew for arid agriculture?' *Atmospheric Research* 87(3–4): 369–76 <https://doi.org/10.1016/j.atmosres.2007.11.016>.

Hadid, A. and Toknok, B. (2020) 'Vegetable and food crop production with microclimate modification', *Journal of Physics: Conference Series* 1434(1): 012027 <https://doi.org/10.1088/1742-6596/1434/1/012027>.

HELVETAS Swiss Intercooperation, Aidenvironment, and MetaMeta (2015) *Water Use Master Plan + 3R* [online] <https://www.academia.edu/43112050/Water_Use_Master_Plan_3R_Facilitators_Manual>. [accessed 21 July 2022].

Herweg, K. and Ludi, E. (1999) 'The performance of selected soil and water conservation measures: case studies from Ethiopia and Eritrea', *Catena* 36(1–2): 99–114 <https://doi.org/10.1016/S0341-8162(99)00004-1>.

Hoekstra, A.Y. and Mekonnen, M.M. (2012) 'The water footprint of humanity', *Proceedings of the National Academy of Sciences* 109(9): 3232–7 <https://doi.org/10.1073/pnas.1109936109>.

Holtslag, A.A.M. (1984) 'Estimation of diabatic wind speed profiles from near-surface weather observations', *Boundary Layer Meteorology* 29: 225–50 <https://doi.org/10.1007/BF00119790>.

Holzer, S. (2012) *Desert or Paradise*, Permanent Publications, New York.

Hooker, J., Duveiller, G. and Cescatti, A. (2018) 'A global dataset of air temperature derived from satellite remote sensing and weather stations', *Scientific Data* 5(1): 1–11 <https://doi.org/10.1038/sdata.2018.246>.

Huang, R., Huang, J-X., Zhang, C., Ma, H-Y., Zhuo, W., Chen, Y.Y., Zhu, D.H., Wu, Q. and Mansaray, L.R. (2020) 'Soil temperature estimation at different depths, using remotely-sensed data', *Journal of Integrative Agriculture* 19(1): 277–90 <https://doi.org/10.1016/S2095-3119(19)62657-2>.

Hupfer, P. (1989) 'Klima im mesoräumigen Bereich', *Abh Meteorol Dienstes DDR* 141: 181–92.

Hussain, Z. and Irfan, M. (2012) 'Sustainable land management to combat desertification in Pakistan', *Journal of Arid Land Studies* 22: 127–9.

Hutjes, R.W.A. (1996) *Transformation of near-surface meteorology in a small-scale landscape with forests and arable land*, PhD thesis, University of Groningen, the Netherlands.

Intergovernmental Panel on Climate Change (IPCC) (2021) 'Summary for policymakers', in V. Masson-Delmotte, P. Zhai, A. Pirani, S.L. Connors, C. Péan, S. Berger, N. Caud, Y. Chen, L. Goldfarb, M.I. Gomis, M. Huang, K. Leitzell, E. Lonnoy, J.B.R. Matthews, T.K. Maycock, T. Waterfield, O. Yelekçi, R. Yu, and B. Zhou (eds), *Climate Change 2021: The Physical Science Basis. Contribution of Working Group I to the Sixth Assessment Report of the Intergovernmental Panel on Climate Change*, Cambridge University Press, Cambridge.

Ismangil, D., Wiegant, D., Yazew, E., van Steenbergen., F., Kool, M., Sambalino, F., Castelli, G. and Bresci, E. (2016) *Managing the Microclimate: Flood-Based Livelihood Network - Practical Note 27* [online] <https://doi.org/10.13140/RG.2.2.15110.78409> [accessed 21 July 2022].

Jackson, R.B., Randerson, J.T., Canadell, J.G. Anderson, R.G., Avissar, R., Baldocchi, D.D., Bonan, G.B., Caldeira, K., Diffenbaugh, N.S., Field, C.B., Hungate, B.A., Jobbágy, E.G., Kueppers, L.M., Nosetto, M.D. and Pataki, D.E. (2008) 'Protecting climate with forests', *Environmental Research Letters* 3(4): 044006 <https://doi.org/10.1088/1748-9326/3/4/044006>.

Jensen, M. (1954) *Shelter Effect: Investigations into the Aerodynamics of Shelter and its Effects in Climate and Crops*, Danish Technical Press, Copenhagen.

Jeon, W.T., Kim, M.T., Seong, K.Y., Lee, J.K., Oh, I.S. and Park, S.T. (2008) 'Changes of soil properties and temperature by green manure under rice-based cropping system', *Korean Journal of Crop Science* 53(4): 413–6.

Jha, S. (2021) 'Over 20 years, these two villages in MP managed to create lush woods on barren land', 101Reporters [website] <https://101reporters.com/article/The_Promise_Of_Commons/Over_20_years_these_two_villages_in_MP_managed_to_create_lush_woods_on_barren_land> [accessed 21 July 2022].

Johnson, M.D. and Lowery, B. (1985) 'Effect of three conservation tillage practices on soil temperature and thermal properties', *Soil Science Society of America Journal* 49(6): 1547–52 <https://doi.org/10.2136/sssaj1985.03615995004900060043x>.

Jones, H.G. (2014) *Plants and Microclimate: A Quantitative Approach to Environmental Plant Physiology*, 3rd edn, Cambridge University Press, Cambridge.

Kader, M.A., Senge, M., Mojid, M.A. and Nakamura, K. (2017) 'Mulching type-induced soil moisture and temperature regimes and water use efficiency of soybean under rain-fed condition in central Japan', *International Soil and Water Conservation Research* 5(4): 302–8 <https://doi.org/10.1016/j.iswcr.2017.08.001>.

Kalma, J.D., McVicar, T.R. and McCabe, M.F. (2008) 'Estimating land surface evaporation: a review of methods using remotely sensed surface temperature data', *Surveys in Geophysics* 29(4): 421–69 <https://doi.org/10.1007/s10712-008-9037-z>.

Kašpar, V., Hederová, L., Macek, M., Müllerová, J., Prošek, J., Surový, P., Wild, J. and Kopecký, M. (2021) 'Temperature buffering in temperate forests: comparing microclimate models based on ground measurements with active and passive remote sensing', *Remote Sensing of Environment* 263: 112522 <https://doi.org/10.1016/j.rse.2021.112522>.

Kaushal, R., Kumar, A., Patra, S., Islam, S., Tomar, J.M.S., Singh, D.V., Mandal, D., Rajkumar, Mehta, H., Chaturvedi, O.P. and Durai, J. (2021) 'In-situ soil moisture conservation in bamboos for the rehabilitation of degraded lands in the Himalayan foothills', *Ecological Engineering* 173: 106437 <https://doi.org/10.1016/j.ecoleng.2021.106437>.

Keating, B.A. and Carberry, P.S. (1993) 'Resource capture and use in intercropping: solar radiation', *Field Crops Research* 34(3–4): 273–301 <https://doi.org/10.1016/0378-4290(93)90118-7>.

Kingra, P. and Kaur, H. (2017) 'Microclimatic modifications to manage extreme weather vulnerability and climatic risks in crop production', *Journal of Agricultural Physics* 17(1): 1–15.

Knörzer, H., Graeff-Hönninger, S., Guo, B., Wang, P. and Claupein, W. (2009) 'The rediscovery of intercropping in China: a traditional cropping system for future Chinese agriculture – a review', in E. Lichtfouse (ed.), *Climate Change, Intercropping, Pest Control and Beneficial Microorganisms. Sustainable Agriculture Reviews*, vol 2, pp. 13–44, Springer, Dordrecht <https://doi.org/10.1007/978-90-481-2716-0_3>.

Kool, M., Mehari Haile, A., Nawaz, K. and van Steenbergen, F. (2016) 'King capillary: the miracle water buffer', TheWaterChannel, 12 December [blog] <https://thewaterchannel.tv/thewaterblog/king-capillary-the-miracle-water-buffer/> [accessed 21 July 2022].

Kuzucu, M. (2019) 'Effects of water harvesting and organic fertilizer on vineyard (*Vitis vinifera* L.) yield and soil moisture content under arid conditions', *Bangladesh Journal of Botany* 48(4): 1115–24 <https://doi.org/10.3329/bjb.v48i4.49067>.

Laipelt, L., Kayser, R.H.B., Fleischmann, A.S., Ruhoff, A., Bastiaanssen, W., Erickson, T.A. and Melton, F. (2021) 'Long-term monitoring of evapotranspiration using the SEBAL algorithm and Google Earth Engine cloud computing', *ISPRS Journal of Photogrammetry and Remote Sensing* 178: 81–96 <https://doi.org/10.1016/j.isprsjprs.2021.05.018>.

Lambert, J.H. (1760) *Photometria*. Augustae Vindelicorum.

Lawston, P.M., Santanello Jr, J.A., Hanson, B. and Arsensault, K. (2020) 'Impacts of irrigation on summertime temperatures in the Pacific Northwest', *Earth Interactions* 24(1): 1–26 <https://doi.org/10.1175/EI-D-19-0015.1>.

Lejeune, Q., Davin, E.L., Gudmundsson, L., Winckler, J. and Seneviratne, S.I. (2018) 'Historical deforestation locally increased the intensity of hot days in northern mid-latitudes', *Nature Climate Change Letters* 8: 386–90 <https://doi.org/10.1038/s41558-018-0131-z>.

Lembrechts, J.J. and Lenoir, J. (2020) 'Microclimatic conditions anywhere at any time!' *Global Change Biology* 26(2): 337–9 <https://doi.org/10.1111/gcb.14942>.

Lembrechts, J.J., Lenoir, J., Scheffers, B.R. and De Frenne, P. (2021a) 'Designing countrywide and regional microclimate networks', *Global Ecology and Biogeography* 30(6): 1168–74 <https://doi.org/10.1111/geb.13290>.
Lembrechts, J.J., van den Hoogen, J., Aalto, J., Ashcroft, M.B., De Frenne, P., Kemppinen, J., Kopecký, M., Luoto, M., Maclean, I.M.D., Crowther, T.W., Bailey, J.J., Haesen, S., Klinges, D.H., Niittynen, P., Scheffers, B.R., Van Meerbeek, K., Aartsma, P., Abdalaze, O., Abedi, M., ... Lenoir, J. (2021b) 'Global maps of soil temperature', *Global Change Biology* 28(9): 3110–44 <https://doi.org/10.1111/gcb.16060>.
Lhomme, J.P. and Vacher, J.J. (2003) 'La mitigación de heladas en los camellones del altiplano andino', *Bulletin de l'Institut Français d'Études Andines* 32(2): 377–99 <https://doi.org/10.4000/bifea.6556>.
Liaghat, S. and Balasundram, S.K. (2010) 'A review: the role of remote sensing in precision agriculture', *American Journal of Agricultural and Biological Sciences* 5(1): 50–55 <https://doi.org/10.3844/ajabssp.2010.50.55>.
Lim, N. and Mukherjee, M. (2021) 'Remote control farming', TheWaterChannel, 16 February [blog] <https://thewaterchannel.tv/thewaterblog/remote-control-farming/> [accessed 7 February 2022].
Lin, B.B. (2007) 'Agroforestry management as an adaptive strategy against potential microclimate extremes in coffee agriculture', *Agricultural and Forest Meteorology* 144(1–2): 85–94 <https://doi.org/10.1016/j.agrformet.2006.12.009>.
Lin, H., Tu, C., Fang, J., Gioli, B., Loubet, B., Gruening, C., Zhou, G., Beringer, J., Huang, J. Dušek, J., Liddell, M., Buysse, P., Shi, P., Song, Q., Han, S., Magliulo, V., Yignian, L. and Grace, J. (2020) 'Forests buffer thermal fluctuation better than non-forests', *Agricultural and Forest Meteorology* 288: 107994 <https://doi.org/10.1016/j.agrformet.2020.107994>.
Locatelli, B., Fedele, G., Fayolle, A. and Baglee, A. (2016) 'Synergies between adaptation and mitigation in climate change finance', *International Journal of Climate Change Strategies and Management* 8: 112–28 <http://dx.doi.org/10.1108/IJCCSM-07-2014-0088>.
Lombardo, U., Canal-Beeby, E., Fehr, S. and Veit, H. (2011) 'Raised fields in the Bolivian Amazonia: a prehistoric green revolution or a flood risk mitigation strategy?' *Journal of Archaeological Science* 38(3): 502–12 <https://doi.org/10.1016/j.jas.2010.09.022>.
Louka, P., Papanikolaou, I., Petropoulos, G.P., Kalogeropoulos, K. and Stathopoulos, N. (2020) 'Identifying spatially correlated patterns between surface water and frost risk using EO data and geospatial indices', *Water* 12(3): 700 <https://doi.org/10.3390/w12030700>.

Maclean, I.M. (2019) 'Predicting future climate at high spatial and temporal resolution', *Global Change Biology* 26(2): 1003–11 <https://doi.org/10.1111/gcb.14876>.

Maclean, I.M., Mosedale, J.R. and Bennie, J.J. (2019) 'Microclima: An R package for modelling meso-and microclimate', *Methods in Ecology and Evolution* 10(2): 280–90 <https://doi.org/10.1111/2041-210X.13093>.

Mahmood, A., Hu, Y., Tanny, J. and Asante, E.A. (2018) 'Effects of shading and insect-proof screens on crop microclimate and production: a review of recent advances', *Scientia Horticulturae* 241: 241–51 <https://doi.org/10.1016/j.scienta.2018.06.078>.

Makarieva, A.M. and Gorshkovak, V.G. (2006) 'Biotic pump of atmospheric moisture as driver of the hydrological cycle on land', *Hydrology and Earth System Sciences Discussions* 3: 2621–73 <https://doi.org/10.5194/hessd-3-2621-2006>.

Manzoni, S., Schimel, J.P. and Porporato, A. (2012) 'Responses of soil microbial communities to water stress: results from a meta-analysis', *Ecology* 93(4): 930–38 <https://doi.org/10.1890/11-0026.1>.

Marshall, J.K. (1967) 'The effect of shelter on the productivity of grasslands and field crops', *Field Crop Abstracts* 20: 1–14.

Mati, B.M. (2006) *Overview of Water and Soil Nutrient Management under Smallholder Rain-Fed Agriculture in East Africa*, Working Paper 105, International Water Management Institute, Colombo, Sri Lanka.

McNamara, K.E. and Buggy, L. (2017) 'Community-based climate change adaptation: a review of academic literature', *Local Environment* 22(4): 443–60 <https://doi.org/10.1080/13549839.2016.1216954>.

Meisner, A., Rousk, J. and Bååth, E. (2015) 'Prolonged drought changes the bacterial growth response to rewetting', *Soil Biology and Biochemistry* 88: 314–22 <https://doi.org/10.1016/j.soilbio.2015.06.002>.

Mekonen, K. and Tesfahunegn, G.B. (2011) 'Impact assessment of soil and water conservation measures at Medego watershed in Tigray, northern Ethiopia', *Maejo International Journal of Science and Technology* 5(3): 312–30.

Mesfin, S., Almeida Oliveira, L.A., Yazew, E., Bresci, E. and Castelli, G. (2019) 'Spatial variability of soil moisture in newly implemented agricultural bench terraces in the Ethiopian plateau', *Water* 11(10): 2134 <https://doi.org/10.3390/w11102134>.

Meyer, S., Bright, R.M., Fischer, D., Schulz, H. and Glaser, B. (2012) 'Albedo impact on the suitability of biochar systems to mitigate global warming', *Environmental Science & Technology* 46(22): 12726–34 <https://doi.org/10.1021/es302302g>.

Milkias, A., Tadesse, T. and Zeleke, H. (2018) 'Evaluating the effects of in-situ rainwater harvesting techniques on soil moisture conservation and grain yield of maize (*Zea mays* L.) in Fedis district, Eastern Hararghe, Ethiopia', *Turkish Journal of Agriculture-Food Science and Technology* 6(9): 1129–33 <https://doi.org/10.24925/turjaf.v6i9.1129-1133.1839>.

Ministry Agriculture Ethiopia (2005) 'Ministry Agriculture Ethiopia Community Based Watershed Management Guideline 2005' [website] <https://wocatpedia.net/wiki/File:Ministry_Agriculture_Ethiopia_Community_Based_Watershed_Management_Guideline_2005_Part_1 _B_a.pdf> [accessed 22 July 2022].

Moene, A.F. and Van Dam, J.C. (2014) *Transport in the Atmosphere-Vegetation-Soil Continuum*, Cambridge University Press, Cambridge.

Monteith, J.L. (1957) 'Dew', *Quarterly Journal of the Royal Meteorological Society* 83(357): 322–41.

Moreno, M.M. and Moreno, A. (2008) 'Effect of different biodegradable and polyethylene mulches on soil properties and production in a tomato crop', *Scientia Horticulturae* 116(3): 256–63 <https://doi.org/10.1016/j.scienta.2008.01.007>.

Muñoz-Romero, V., Lopez-Bellido, L. and Lopez-Bellido, R.J. (2015) 'Effect of tillage system on soil temperature in a rainfed Mediterranean Vertisol', *International Agrophysics* 29: 467–73 <https://doi.org/10.1515/intag-2015-0052>.

Nature^Squared, SamSamWater, and Justdiggit (no date) GreenerLAND [website] <https://www.greener.land> [accessed 14 September 2021].

Nedjai, R., Azaroual, A., Chlif, K., Bensaid, A., Al-Sayah, M. and Ysbaa, L. (2018) 'Impact of ponds on local climate: a remote sensing and GIS application contribution to ponds of Brenne (France)', *Journal of Earth Science and Climate Change* 9(12) <https://doi.org/10.4172/2471-8556.1000503>.

Norman, L.M., Brinkerhoff, F., Gwilliam, E., Guertin, D.P., Callegary, J., Goodrich, D.C., Nagler, P.L. and Gray, F. (2015) 'Hydrologic response of streams restored with check dams in the Chiricahua Mountains, Arizona', *River Research and Applications* 32(4): 519–27 <https://doi.org/10.1002/rra.2895>.

Norris, C., Hobson, P. and Ibisch, P.L. (2011) 'Microclimate and vegetation function as indicators of forest thermodynamic efficiency', *Journal of Applied Ecology* 49(3): 562–70 <https://doi.org/10.1111/j.1365-2664.2011.02084.x>.

Nyawade, S.O., Karanja, N.N., Gachene, C.K.K., Gitari, H.I., Schulte-Geldeman, E. and Parker, M. (2019) 'Intercropping optimizes soil temperature and increases crop water productivity and radiation use efficiency of rainfed potato', *American Journal of Potato Research* 96: 457–71 <https://doi.org/10.1007/s12230-019-09737-4>.

Nyssen, J., Poesen, J., Gebremichael, D., Vancampenhout, K., Dáes, M., Yihdego, G., Govers, G., Leirs, H., Moeyersons, J., Naudts, J., Haregeweyn, N., Haile, M. and Deckers, J. (2007) 'Interdisciplinary on-site evaluation of stone bunds to control soil erosion on cropland in Northern Ethiopia', *Soil and Tillage Research* 95: 151–63 <https://doi.org/10.1016/j.still.2006.07.011>.

Nyssen, J., Clymans, W., Descheemaeker, K., Poesen, J., Vandecasteele, I., Vanmaercke, M., Zenebe, A., Van Camp, M., Haile, M., Haregeweyn, N., Moeyersons, J., Martens, K., Gebreyohannes, T., Deckers, J. and Walraevens, K. (2010) 'Impact of soil and water conservation measures on catchment hydrological response: a case in north Ethiopia', *Hydrological Processes* 24(13): 1880–95 <https://doi.org/10.1002/hyp.7628>.

Oke, T.R. (1995) *Boundary Layer Climates*, 2nd edn, Routledge.

Onwuka, B. and Mang, B. (2018) 'Effects of soil temperature on some soil properties and plant growth', *Advances in Plants & Agriculture Research* 8(1): 34–7 <https://doi.org/10.15406/apar.2018.08.00288>.

Patil, Y.M., Belgaumi, M.I., Maurya, N.L., Kusad, V.S., Mansur, C.P. and Patil, S.L. (1995) 'Impact of mechanical and vegetative barriers on soil and moisture conservation', *Indian Journal of Soil Conservation* 23(3): 254–5.

Peterson, C.A., Eviner, V.T. and Gaudin, A.C. (2018) 'Ways forward for resilience research in agroecosystems', *Agricultural Systems* 162: 19–27 <https://doi.org/10.1016/j.agsy.2018.01.011>.

Placidi, P., Morbidelli, R., Fortunati, D., Papini, N., Gobbi, F. and Scorzoni, A. (2021) 'Monitoring soil and ambient parameters in the IoT precision agriculture scenario: an original modeling approach dedicated to low-cost soil water content sensors', *Sensors* 21(15): 5110 <https://doi.org/10.3390/s21155110>.

Rashid, M., Alvi, S., Kausar, R. and Akram, M.I. (2016) 'The effectiveness of soil and water conservation terrace structures for improvement of crops and soil productivity in rainfed terraced system', *Pakistan Journal of Agricultural Sciences* 53(1).

Rawat, K.S., Singh, S.K. and Pal, R.K. (2019) 'Synergetic methodology for estimation of soil moisture over agricultural area using Landsat-8 and Sentinel-1 satellite data', *Remote Sensing Applications: Society and Environment* 15: 100250 <https://doi.org/10.1016/j.rsase.2019.100250>.

Reichenau, T.G., Korres, W., Montzka, C., Fiener, P., Wilken, F., Stadler, A., Waldhoff, G. and Schneider, K. (2016) 'Spatial heterogeneity of leaf area index (LAI) and its temporal course on arable land: combining field measurements, remote sensing and simulation in a comprehensive data analysis approach (CDAA)', *PLoS ONE* 11(7): e0158451 <https://doi.org/10.1371/journal.pone.0158451>.

Rodrigo, V.H.L., Iqbal, S.M.M., Munasinghe, E.S., Balasooriya, B.M.D.C. and Jayathilake, P.M.M. (2014) 'Rubber in East assures the perceived benefits; a case study showing increased rubber production, amelioration of the climate and improved rural livelihood', *Journal of the Rubber Research Institute of Sri Lanka* 94: 33–42.

Roldán, J., Chipana, R., Moreno, M.F., del Pino, J.L. and Bosque, H. (2004) Suka Kollus; tecnología prehispánica de riego y drenaje en proceso de abandono: estrategias mixtas de diseño y manejo, Universidad de Córdoba, Spain and Universidad de La Paz, Bolivia <http://ceer.isa.utl.pt/cyted/brasil2008/tema5/Sessao%20V_JRoldan_RChipana_Suka%20Kollus.pdf> [accessed 5th May 2022; website no longer active].

Roncoli, C., Orlove, B.S., Kabugo, M.R. and Waiswa, M.M. (2011) 'Cultural styles of participation in farmers' discussions of seasonal climate forecasts in Uganda', *Agriculture and Human Values* 28(1): 123–38 <https://doi.org/10.1007/s10460-010-9257-y>.

Rosenberg, N.J., Blad, B.L. and Verma, S.B. (1983) *Microclimate: The Biological Environment*, John Wiley & Sons, New York.

Sagan, V., Maimaitijiang, M., Sidike, P., Eblimit, K., Peterson, K. T., Hartling, S., Esposito, F., Khanal, K., Newcomb, M., Pauli, D., Ward, R., Fritschi, F., Shakoor, N. and Mockler, T. (2019) 'UAV-based high resolution thermal imaging for vegetation monitoring, and plant phenotyping using ICI 8640 P, FLIR Vue Pro R 640, and thermomap cameras', *Remote Sensing* 11(3): 330 <https://doi.org/10.3390/rs11030330>.

Sandhu, S.K. and Dhaliwal, L.K. (2016) 'Crop geometry effects on relative humidity variation within wheat crop', *Asian Journal of Environmental Science* 11(1): 94–101 <https://doi.org/10.15740/HAS/AJES/11.1/94-101>.

Schultze, S.R., Campbell, M.N., Walley, S., Pfeiffer, K. and Wilkins, B. (2021) 'Exploration of sub-field microclimates and winter temperatures: implications for precision agriculture', *International Journal of Biometeorology* 65: 1043–52 <https://doi.org/10.1007/s00484-021-02086-0>.

Schwartz, J.D. (2013) 'Clearing forests may transform local—and global—climate', *Scientific American*, 4 March [online] <https://www.scientificamerican.com/article/clearing-forests-may-transform-local-and-global-climate/> [accessed 22 July 2022]

Schwartz, J.D. (2020) *The Reindeer Chronicles: And Other Inspiring Stories of Working with Nature to Heal the Earth*, Chelsea Green Publishing, White River Junction, VT.

Schwingshackl, C., Hirschi, M. and Seneviratne, S.I. (2017) 'Quantifying spatiotemporal variations of soil moisture control on surface energy balance and near-surface air temperature', *Journal of Climate* 30(18): 7105–24 <https://doi.org/10.1175/JCLI-D-16-0727.1>.

Scott, D.C. (2000) *Soil Physics, Agricultural and Environmental Applications*, Iowa State University Press, Ames, IA.

Semenzato, R., Falciai, M. and Bresci, E. (1998) 'The project "Fog as a new water resource for the sustainable development of the ecosystems of the Peruvian and Chilean coastal desert"', *First International Conference on Fog and Fog Collection, Vancouver (Canada), 19–24 July 1998*, pp. 457–60.

Shahak, Y. (2008) 'Photo-selective netting for improved performance of horticultural crops: a review of ornamental and vegetable studies carried out in Israel', *Acta Horticulturae* 770: 161–8 <https://doi.org/10.17660/ActaHortic.2008.770.18>.

Sharpe, D.M. (1987) 'Microclimatology', in: *Climatology: Encyclopedia of Earth Science*, pp. 572–81, Springer, Boston, MA.

Sharratt, B.S. and McWilliams, D.A. (2005) 'Microclimatic and rooting characteristics of narrow-row versus conventional-row corn', *Agronomy Journal* 97(4): 1129–35 <https://doi.org/10.2134/agronj2004.0292>.

Shaxson, F. and Barber, R. (2003) *Optimizing Soil Moisture for Plant Production: The Significance of Soil Porosity* [online], UN-FAO <http://hdl.handle.net/10919/65454> [accessed 28 July 2022].

Sheil, D. (2018) 'Forests, atmospheric water and an uncertain future: the new biology of the global water cycle', *Forest Ecosystems* 5(1): 1–22 <https://doi.org/10.1186/s40663-018-0138-y>.

Shen, Y., Zhao, Z. and Shi, G. (2008) 'The progress in variation of surface solar radiation, factors and probable climatic effects', *Advances in Earth Science* 23(9): 915–24.

Shen, Y., McLaughlin, N., Zhang, X., Xu, M. and Liang, A. (2018) 'Effect of tillage and crop residue on soil temperature following planting for a Black soil in northeast China', *Science Report* 8: 4500 <https://doi.org/10.1038/s41598-018-22822-8>.

Sida, T.S., Baudron, F., Kim, H. and Giller, K.E. (2018) 'Climate-smart agroforestry: *Faidherbia albida* trees buffer wheat against climatic extremes in the Central Rift Valley of Ethiopia', *Agricultural and Forest Meteorology* 248: 339–47 <https://doi.org/10.1016/j.agrformet.2017.10.013>.

Silvestri, M., Romaniello, V., Hook, S., Musacchio, M., Teggi, S. and Buongiorno, M.F. (2020) 'First comparisons of surface temperature estimations between ECOSTRESS, ASTER and Landsat 8 over Italian volcanic and geothermal areas', *Remote Sensing* 12(1): 184 <https://doi.org/10.3390/rs12010184>.

Smith, M. and Steduto, P. (2012) 'Yield response to water: the original FAO water production function', *FAO Irrigation and Drainage Paper* 66: 6–13.

Sobrino, J.A., Julien, Y., Jiménez-Muñoz, J.C., Skokovic, D. and Sòria, G. (2020) 'Near real-time estimation of sea and land surface temperature for MSG SEVIRI sensors', *International Journal of Applied Earth Observation and Geoinformation* 89: 102096 <https://doi.org/10.1016/j.jag.2020.102096>.

Soomro, A.G., Babar, M.M., Arshad, M., Memon, A., Naeem, B. and Ashraf, A. (2020) 'Spatiotemporal variability in spate irrigation systems in Khirthar National Range, Sindh, Pakistan (case study)', *Acta Geophysica* 68(1): 219–28 <https://doi.org/10.1007/s11600-019-00392-1>.

Spracklen, D.V., Arnold, S.R. and Taylor, C.M. (2012) 'Observations of increased tropical rainfall preceded by air passage over forests', *Nature* 489(7415): 282–5 <https://doi.org/10.1038/nature11390>.

Sprenkle-Hyppolite, S.D., Latimer, A.M., Young, T.P. and Rice, K.J. (2016) 'Landscape factors and restoration practices associated with initial reforestation success in Haiti', *Ecological Restoration* 34(4): 306–16 <https://doi.org/10.3368/er.34.4.306>.

Stickler, F.C. and Laude, H.H. (1960) 'Effect of row spacing and plant population on performance of corn, grain sorghum and forage sorghum 1', *Agronomy Journal* 52(5): 275–7 <https://doi.org/10.2134/agronj1960.00021962005200050011x>.

Stigter, C.J. (1984a) 'Mulching as a traditional method of micro-climate management', *Archives for Meteorology, Geophysics, and Bioclimatology, Series B* 35(1–2): 147–54 <https://doi.org/10.1007/BF02269417>.

Stigter, C.J. (1984b) '*Traditional use of shade: a method of microclimate manipulation*', *Archives for Meteorology, Geophysics, and Bioclimatology* 34: 203–10 <https://doi.org/10.1007/BF02275684>.

Stomph, T., Dordas, C., Baranger, A., de Rijk, J., Dong, B., Evers, J., Gu, C., Li, L., Simon, J., Steen Jensen, E., Wang, Q., Wang, Y., Wang, Z., Xu, H., Zhang, C., Zhang, L., Zhang, W., Bedoussac, L. and van der Werf, W. (2019) 'Designing intercrops for high yield, yield stability and efficient use of resources: are there principles? Advances in Agronomy 160: 1–50 <https://doi.org/10.1016/bs.agron.2019.10.002>.

Stoutjesdijk, P. and Barkman, J.J. (1992) *Microclimate, Vegetation and Fauna*, Opulus Press, Uppsala/Leiden.

Stull, R.B. (1988) *An Introduction to Boundary Layer Meteorology*, vol. 13, Springer Science & Business Media, Berlin/Heidelberg.

Sun, D. and Dickinson, G.R. (1994) 'Wind effect on windbreak establishment in Northern Australia', *Tree Planter's Notes* 45: 72–5.

Tanny, J. and Cohen, S. (2003) 'The effect of a small shade net on the properties of wind and selected boundary layer parameters above and within a citrus orchard', *Biosystems Engineering* 84(1): 57–67 <https://doi.org/10.1016/S1537-5110(02)00233-7>.

Tayade, A.S., Geetha, P., Dhanapal, R. and Hari, K. (2016) 'Effect of in situ trash management on sugarcane under wide row planting system', *Journal of Sugarcane Research* 6(1): 35–41.

Taye, G., Poesen, J., Wesemael, B.V., Vanmaercke, M., Teka, D., Deckers, J., Goosse, T., Maetens, W., Nyssen, J. and Hallet, V. (2013) 'Effects of land use, slope gradient, and soil and water conservation structures on runoff and soil loss in semi-arid northern Ethiopia', *Physical Geography* 34: 236–59 <https://doi.org/10.1080/02723646.2013.832098>.

Tengnäs, B. (1994) *Agroforestry Extension Manual for Kenya*, World Agroforestry Centre, Nairobi.

Tonolla, D., Acuna, V., Uehlinger, U., Frank, T. and Tockner, K. (2010) 'Thermal heterogeneity in river floodplains', *Ecosystems* 13(5): 727–40 <https://doi.org/10.1007/s10021-010-9350-5>.

Trilles, S., Juan, P., Díaz-Avalos, C., Ribeiro, S. and Painho, M. (2020) 'Reliability evaluation of the data acquisition potential of a low-cost climatic network for applications in agriculture', *Sensors* 20(22): 6597 <https://doi.org/10.3390/s20226597>.

Tucci, G., Parisi, E.I., Castelli, G., Errico A., Corongiu, M., Sona, G., Viviani, E., Bresci, E. and Preti, F. (2019) 'Multi-Sensor UAV application for thermal analysis on a dry-stone terraced vineyard in rural Tuscany landscape', *ISPRS International Journal of Geo-Information* 8: 87 <https://doi:10.3390/ijgi8020087>.

Unwin, D.M. and Corbet, S.A. (1991) *Insects, Plants, and Microclimate*, Naturalists' Handbooks Series, Band 15, Richmond Publishing, Oxford.

Van De Giesen, N., Hut, R. Andreini, M. and Selker, J.S. (2013) 'Trans-African Hydro-Meteorological Observatory (TAHMO): a network to monitor weather, water, and climate in Africa', *AGU Fall Meeting Abstracts* 2013: H52C-04.

Van Steenbergen, F. (2013a) 'The power of dew', TheWaterChannel, 21 January [blog] <https://thewaterchannel.tv/thewaterblog/the-power-of-dew/> [accessed 28 July 2022].

Van Steenbergen, F. (2013b) 'The coming plastic revolution', TheWater Channel, 4 March [blog] <https://thewaterchannel.tv/thewaterblog/the-coming-plastic-revolution/> [accessed 28 July 2022].

Van Steenbergen, F. and Agujetas, M. (2018) 'Tackling dust', TheWaterChannel, 9 March [blog] <https://thewaterchannel.tv/thewaterblog/tackling-dust/> [accessed 28 July 2022].

Van Steenbergen, F., Tuinhof, A., Knoop, L. and Kauffman, J.H. (2011) *Transforming Landscapes, Transforming Lives: The Business of Sustainable Water Buffer Management*, Wageningen 3R Water.

Van Steenbergen, F., Arroyo-Arroyo, F., Rao, K., Alemayehu Hulluka, T., Woldearegay, K. and Deligianni, A. (2021) *Green Roads for Water: Guidelines for Road Infrastructure in Support of Water Management and Climate Resilience*, World Bank Group, Washington, DC <https://doi.org/10.1596/978-1-4648-1677-2>.

Van Wijk, W.R. (1963) *Physics of Plant Environment*, North-Holland Publishing Company, Amsterdam.

Villani, L., Castelli, G., Sambalino, F., Almeida Oliveira, L. and Bresci, E. (2020) 'Integrating UAV and satellite data to assess the effects of agroforestry on microclimate in Dodoma region, Tanzania', *IEEE International Workshop on Metrology for Agriculture and Forestry (MetroAgriFor)*, pp. 338–42 <https://doi.org/10.1109/MetroAgriFor50201.2020.9277643>.

Villani, L., Castelli, G., Sambalino, F., Almeida Oliveira, L. and Bresci, E. (2021) 'Influence of trees on landscape temperature in semi-arid agro-ecosystems of East Africa', *Biosystems Engineering* 212: 185–99 <https://doi.org/10.1016/j.biosystemseng.2021.10.007>.

Wang, Q., Li, J., Jin, T., Chang, X., Zhu, Y., Li, Y., Sun, J. and Li, D. (2020) 'Comparative analysis of Landsat-8, Sentinel-2, and GF-1 data for retrieving soil moisture over wheat farmlands', *Remote Sensing* 12(17): 2708 <https://doi.org/10.3390/rs12172708>.

Wiegant, D.A. and van Steenbergen, F. (2017) 'Steps towards groundwater-sensitive land-use governance and management practices', in K.G. Villholth, E. Lopez-Gunn, K. Conti, A. Garrido, and J. Van Der Gun (eds), *Advances in Groundwater Governance*, pp. 307–27, Taylor & Francis Group, London.

Wild, J., Kopecký, M., Macek, M., Šanda, M., Jankovec, J. and Haase, T. (2019) 'Climate at ecologically relevant scales: a new temperature and soil moisture logger for long-term microclimate measurement', *Agricultural and Forest Meteorology* 268: 40–47 <https://doi.org/10.1016/j.agrformet.2018.12.018>.

Wilken, G.C. (1972) 'Microclimate management by traditional farmers', *Geographical Review* 62(4): 544–60 <https://doi.org/10.2307/213267>.

Wohlleben, P. (2016) *The Hidden Life of Trees: What They Feel, How They Communicate—Discoveries from a Secret World*, vol. 1, Greystone Books, Vancouver.

Yang, C., Zhan, Q., Lv, Y. and Liu, H. (2019) 'Downscaling land surface temperature using multiscale geographically weighted regression over heterogeneous landscapes in Wuhan, China', *IEEE Journal of Selected Topics in Applied Earth Observations and Remote Sensing* 12(12): 5213–22 <https://doi.org/10.1109/JSTARS.2019.2955551>.

Yang, W.P., Guo, T.C., Liu, S.B., Wang, C.Y., Wang, Y.H. and Ma, D.Y. (2008) 'Effects of row spacing in winter wheat on canopy structure and microclimate in later growth stage', *Chinese Journal of Plant Ecology* 32(2): 485 <https://doi.org/10.3773/j.issn.1005-264x.2008.02.028>.

Yildiz, İ. and Rattan, L.A.L. (1996) 'Effect of row orientation and mulching on soil temperature and moisture regimes', *Turkish Journal of Agriculture and Forestry* 20(4): 319–25 <https://doi.org/10.55730/1300-011X.2907>.

Yoshino, M.M. (1975) *Climate in a Small Area: An Introduction to Local Meteorology*, University of Tokyo Press.

Zapater, M., Hossann, C., Bréda, N., Bréchet, C., Bonal, D. and Granier, A. (2011) 'Evidence of hydraulic lift in a young beech and oak mixed forest using 18 O soil water labelling', *Trees* 25(5): 885–94 <https://doi.org/10.1007/s00468-011-0563-9>.

Zawadzka, J., Corstanje, R., Harris, J. and Truckell, I. (2020) 'Downscaling Landsat-8 land surface temperature maps in diverse urban landscapes using multivariate adaptive regression splines and very high-resolution auxiliary data', *International Journal of Digital Earth* 13(8): 899–914 <https://doi.org/10.1080/17538947.2019.1593527>.

Zellweger, F., De Frenne, P., Lenoir, J., Rocchini, D. and Coomes, D. (2019) 'Advances in microclimate ecology arising from remote sensing', *Trends in Ecology & Evolution* 34(4): 327–41 <https://doi.org/10.1016/j.tree.2018.12.012>.

Zhang, C. and Wen, M. (2014) 'Using satellite data to estimate solar radiation of clear sky over Fujian', *Journal of Natural Resources* 29: 1496–507.

Zhang, Q., Wang, Y., Wu, Y., Wang, X., Du, Z., Liu, X. and Song, J. (2013) 'Effects of biochar amendment on soil thermal conductivity, reflectance, and temperature', *Soil Science Society of America Journal* 77(5): 1478–87 <https://doi.org/10.2136/sssaj2012.0180>.

Zullu, J. Jr, Pinto, H.S., Assa, E.D. and Evangelista, S.R.M. (2008) 'Potential economic impacts of global warming on two Brazilian commodities according to IPCC prognostics', *Terrae* 3(1): 28–39.

Appendix A: Local climate monitoring tools overview

Overview of local climate monitoring tools and their purpose, spatial and temporal resolution, costs, and advantages and disadvantages.

Local climate monitoring tools	Purpose of measurements	Spatial and temporal resolution	Cost	Advantages and disadvantages
Global climate data sets				
Global climate data sets (e.g. GLDAS, ERA5)	A large number of atmospheric, land, and oceanic climate variables.	*Spatial resolution* 10 km × 10 km × 50 km × 50 km *Temporal resolution* 1 hour	Data is freely available.	*Advantages* • Extensive list of measured factors. • Suitable for first level assessments when there is no finer grained information. *Disadvantages* • Predictions based on coarse resolution.
Ground measurements				
Local weather stations	Air temperature Air humidity Atmospheric pressure Light intensity Rainfall Wind speed Wind direction	*Spatial resolution* Point measurement *Temporal resolution* Real-time	Standard stations are cheaper. Automated stations on a specific site are more expensive.	*Advantages* • Displays the real-time situation. *Disadvantages* • Data describes very localized conditions. • Maintenance is not always optimal.

134 MANAGING THE LOCAL CLIMATE

Local climate monitoring tools	Purpose of measurements	Spatial and temporal resolution	Cost	Advantages and disadvantages
				• When not placed at the right spot, the data is not representative of a specific local climate.
Field sensors	Soil temperature	*Spatial resolution*	Cheap to expensive (Depending on the capabilities of the sensor).	*Advantages*
	Air temperature			• Displays the real-time situation.
	Soil moisture	Point measurement		
	Soil temperature	*Temporal resolution*		• Highly flexible in use.
	Solar radiation	Real-time		• Highly localized measurements.
	Leaf temperature			
	Thermal radiation			*Disadvantages*
	Atmospheric pressure			• Requires skills to set up a system.
	CO_2 levels			• High risk of damage during operation.
Unmanned aerial vehicle (UAV) remote sensing				
Thermal cameras on lightweight drones	Surface temperature	*Spatial resolution*	High	*Advantages*
	Detailed vegetation structure	Very fine		• Extraordinary quality.
	Data can be used to derive local climate variables, such as air temperature, soil moisture, soil temperature, and air humidity (examples have been given throughout the text).	*Temporal resolution* Intermittent		• High resolution, scale-appropriate for microclimate data.
				Disadvantages
				• Intermittent measurements.
				• Limited areal coverage.
Satellite remote sensing				
Geostationary satellites (e.g. MSG, GOES)	Solar radiation	*Spatial resolution* 2 to 3 km	Data is freely available.	*Advantages* • Easy access.

Local climate monitoring tools	Purpose of measurements	Spatial and temporal resolution	Cost	Advantages and disadvantages
	Data from different bands can be used to derive local climate variables, such as air temperature, soil moisture, soil temperature, and air humidity (examples have been given throughout the text).	Temporal resolution 5 to 15 minutes		Disadvantages • Not all local climate variables are directly measured. For this, data processing and calculations are required. • Only representative of the mesoclimate, not the microclimate.
Landsat-8	Data from different bands can be used to derive local climate variables, such as air temperature, soil moisture, soil temperature, and air humidity (examples have been given throughout the text).	Spatial resolution Measures land surface temperature at a resolution of 100 metres. Temporal resolution 8 days	Data is freely available.	Advantages • Ease of use. • Long time series. Disadvantages • Not all local climate variables are directly measured. For this, data processing and calculations are required. • Medium temporal resolution.
Sentinel-3	Data from different bands can be used to derive local climate variables, such as air temperature, soil moisture, soil temperature, and air humidity (examples have been given throughout the text).	Spatial resolution Measures land surface temperature at a resolution of 1,000 metres.	Data is freely available.	Advantages • Easy access.

Local climate monitoring tools	Purpose of measurements	Spatial and temporal resolution	Cost	Advantages and disadvantages
		Temporal resolution 1 day		*Disadvantages* • Not all local climate variables are directly measured. For this, data processing and calculations are required.
MODIS	Data from different bands can be used to derive local climate variables, such as air temperature, soil moisture, soil temperature, and air humidity (examples have been given throughout the text).	*Spatial resolution* Measures land surface temperature at a resolution of 1,000 metres. *Temporal resolution* 1 day	Data is freely available.	*Advantages* • Easy access. • Extensive time-series available. *Disadvantages* • Not all local climate variables are directly measured. For this, data processing and calculations are required.
Planet Labs	Data from different bands can be used to derive local climate variables, such as air temperature, soil moisture, soil temperature, and air humidity (examples have been given throughout the text).	*Spatial resolution* 3 to 5 metres (depending on the sensor considered). *Temporal resolution* Orbits the planet every 90 minutes, providing near real-time images for time-sensitive monitoring.	A subscription is required to access the data.	*Advantages* • High spatial and temporal resolution. *Disadvantages* • Data is not freely available. • Not all local climate variables are directly measured. For this, data processing and calculations are required.

Local climate monitoring tools	*Purpose of measurements*	*Spatial and temporal resolution*	*Cost*	*Advantages and disadvantages*
ECOSTRESS sensor	Land surface temperature Plant temperature	*Spatial resolution* 38 to 69 metres. *Temporal resolution* Daily	Data is freely available.	Advantages • Provides insights into climatic effects on crops. Disadvantages • Not all local climate variables are directly measured. For this, data processing and calculations are required.

www.ingramcontent.com/pod-product-compliance
Ingram Content Group UK Ltd.
Pitfield, Milton Keynes, MK11 3LW, UK
UKHW021831210426
5322IPUK00004B/128